ACCORDANCE

ACCORDANCE

A Study of Consciousness

ALFRED JOHN DALRYMPLE

ISBN 978-0-9846242-7-0
Library of Congress Catalog Card Number:
2014914097

This book has my four essays on consciousness:
"I"
Awareness as the Ground of Being
As Time Goes By
Consciousness as the Primary Field

Published by Dart Books
P.O. Box 744, Unalaska, AK 99685
907-581-3701
Or
P.O. Box 149, Northwood, N.H. 03261
603-942-5993

ACCORDANCE

Foreword (A)

The four parts of this book are my essays about consciousness. The last two have a conversational manner, partly because I feel comfortable having a friend move me along by asking questions.

Each part considers that we must try to know time and distance as one thing following another, related to the speed of light...but, also, as pertaining to the "connectedness" of a reality larger than our universe.

The larger reality, I call "Our Something", and I believe its "enabler" is Consciousness. It seems to me, then, that time and distance is also, and primarily, consciousness moving itself forward.

Consciousness is immediate. Considering this...we can, perhaps, have clearer understanding of cause and effect, the separable nature of things, freewill, and fate.

...

In this universe, a particle and its anti-particle collide and annihilate one another...in action faster than the speed of light. It was suggested by Richard Feynman, in support of

Einstein's thought, that during this action there is a move backward in time.

In action which resembles particle to particle encounter… Consciousness is with each unit of itself, in movement which is "immediate". It's likely, then, that during the action there is enablement by way of a move back in time…and ahead. I see a resultant connection between the particle's past, present (the "now" of it),

ACCORDANCE

Foreword (B)

and its future."Our Something" is effected onward in each unit... continually the "same as" and, so, connected...but also "different from" and, so, separate.

• • •

The book's first part..."I"...says Consciousness underlies our existence. My presentation of this leads to defense of a certainty that Consciousness precedes DNA, and not the other way around.

This is true for all things.

As for man...I have some thoughts about freewill.

• • •

In the second part I address, among other things, "entanglement"...wherein separated bits of matter, even if greatly distanced, co-respond if you touch one. For those who believe nothing of this action can exceed the speed of light, there is no message sent between the bits. This erases the separable nature of things, and kills cause and effect. It says you and I are without freewill.

It says we're robots.

I say there is message sent. Time and distance is relative... also, and primarily...to the immediacy of Consciousness.

In the third part...I converse with my friend Nyellie Trululu Barton Jangmu Sherpa. Nell Barton, for short.

When we use the word love...in debate...and call it

ACCORDANCE

Foreword (C)

necessary imperfection, we agree that nothing of "Our Something" would exist without it.

We speak of freewill...agreeing that the human "I" can make choices alone, as separate from genes, other men, or the "I" of the Entirety.

• • •

In the fourth part...as to Consciousness...the words "enable" and "encounter" gain prominence.

As for freewill and fate, I arrive at greater certainty that both are true. I try to state why this is so.

"I"

Alfred John Dalrymple

"I"

Foreword (A)

Some large questions are bringing us to whistling in the dark... the answers wanting and not wanting to be heard.

I heard a physicist say "I have a problem with "Nothing". Then he was gone...with the wind.

• • •

Of space...where "Nothing" seems to be there...why is there "Something" rather than nothing?

• • •

If there is a Ground of Being, of our existence...it rests in its own fullness as one thing, yet it has infinite parts able to manifest into a reality demanding "twoness" and "symmetry". Is there any other ground, then, but Sentience, Awareness, Consciousness?

How could "Our Something", or an "Entirety", be...this necessitating physical laws for the continuance of it...unless it knew it existed and made, or allowed, the laws.

• • •

I feel giddy…talking about things I can't answer. Should I, too, let the wind blow…and be gone with it? But…wait! I think the physicist and I can let our imagi-nation loose, even sober. Sure!

Most of mankind has unquestioned certainty about beginnings and endings and purposes. I speak of Insti-tutional thought…the kind that has caused almost all wars, slaughters, and imprisonments of body and soul.

We must speak our minds in order to oppose, to offer
Alternatives to the blinding certainty.

Foreword (B)

I use my imagination with sincerity. If the physicist and you and I study the "stuff" we're made of, and wonder at the symmetry, hidden or not, that holds it together, we can see that not only are we all equal in the wilderness...we have a chance to be harmonic here.

"I am better than you!"...the thread running through much of "church" (not through religion itself)...is a blindness needing to be fixed.

• • •

In my book, please see question marks at every turn, acknowledging human error. I welcome disagreementfrom readers who know they can be wrong.

Institutional certainties have, for many, effectively destroyed "wondering about"...or that which allows error and the need to alter one's thought. They've nearly robotized man, killing questions of mortality, and the fear of standing alone and separate in choices...a condition necessary to being human.

Institutional sheepfold domination, able to almost destroy freewill...is a foolish consistency which has led millions into darkness.

This is the time for those with integrity, who think they can be wrong...being human...to step forward into difficult places and say what they believe.

• • •

To understand Consciousness I used my thought, as imagination, even beyond Dark Matter and Dark Energy…

whatever they are…to the beginning of things…whatever that is.

I came to a bridge called "imagination"…and this announcement:

Foreword (C)

For centuries…as to how it all began…
few men have crossed this way.
If you, too, sow unquestioned certainty…
I'll spit out you…like a grape seed.

"I"

Alfred John Dalrymple

Why is there "Something" rather than "Nothing"?
And...why are the laws of "Something" harmonic?
Made of the stuff...
man has "I".
Why?

Perhaps...

"Something rather than Nothing" didn't begin with a singularity. "Something" is throughout its own content.

Our singularity entered...returning to "Something". And, likely, there are now, and have been, many universes other than ours, each formed as to their own singularity...they, too, being part of "Something".

Rejecting the logic that in a place assumed to be empty a thing can appear to us and then disappear, it is accepted by me that, in this universe and its environs, there is no emptiness allowing matter to come from.

All so-called space is filled.

"I"

Part One

A.
Sentience resides in all things.

By sentience I mean the awareness available to a thing... either as "I" or part of it, as partaking of the laws necessary to existence.

Consciousness, is an ethereal materiality, in the sense of being able to touch, and direct, and make the material world.

All things have universal or "pure" consciousness, acting and/or maintaining as part of both "I" and the necessary physical laws.

Universal awareness, pure consciousness, is in all things, including the cells and the "I" of man.

The entirety of man's mind has "I"...but the parts of it maintaining and acting according to the laws...as within this "I", this particular entirety.

B.
Whatever mass is...I believe all physical laws and the building blocks they allow are harmonious...or the "Something" would not have continued.

Existence loved itself, in the sense of wanting tocontinue, thus allowing for growth, for becoming "best existence", within its laws.

...

It seems likely there was a time, during non-zero temperature, in which Consciousness radiated the conditions of materiality...later resulting in fields, either strings, or particles able to house electromagnetic influences, this then onward to atoms. But...whatever the process...laws allowing materiality presented un-conscious purpose to all parts. Then...parts maintained themselves according to the laws.

...

There was a focal point of awareness, a realization of existence, not anywhere in particular, but everywhere "here". The time and distance of it relating to the "immediacy" of consciousness.

Is the "I" of man embodied in a particular part of the head? It's not anywhere in particular, but is everywhere in particular. "I am here"...everywhere here". The indi-vidual parts obey the laws of existence...they have un-conscious purpose...and are touched by the "I" of this place.

...

If the first ethereality can be partly defined as "Possibility"...it would indicate that Consciousness' parts had pure consciousness, as to purpose, which was to exist.

...

Later, within a cosmos of universes, "Purpose" was able to be perceived as embodied in a centrality, or focal point...or "God". This understanding would be within things that could directly share "I". One of these sharers is man.

(the above use of the word "later" refers to the likeli-hood that "possibility", itself, is centered in the meaning of consciousness. And so, it refers to all exist-ence. This as to one thing following another, relative to the speed of light...and, also, as to a carpet-like time, where immediacy is a "throughoutness" on the carpet)

• • •

The laws allowing "Something" according to that which maintains itself, probably are essentially the same everywhere, this including other universes and dimensions. Anything in them with the capacity to house the "stuff", as appropriate to having thought in large measure, will have "I".

C.

For man's present thought, and study, the simplest form of consciousness is the self-maintenance of the atom and its parts.

In the atom, parts acting in accord with the laws of electromagnetism, for instance, are aware as to the other parts, but only as part of universal "I". The parts of the atom don't have "I", but they have the purpose of it...that is, they act not only to exist but to continue to do so.

• • •

The self-maintenance of any building block...atom, string, or other unit...has rhythmic movement, or it couldn't maintain.

Without harmony, in "possibility", existence would not have arrived at an "I" wondering about it.

• • •

Existence immediately loved itself, and knew it did, and knew it had to in order to persist. It harmonically moved... movement being a key to existence. (this in non-zero temperature, offering nearly free flow of whatever radiates) Then... the forces of nature were determined by immediate introduction of "twoness" andconsequent symmetry within radiated electromagnetic and other forces.

D.

Earth's inanimate and animate entities are made ofthe stuff of the stars. They continue onward...knowing how to maintain, and, in some cases, both animate and inanimate, how to improve, or try to.

Some of the animate entities have a house, or skull, containing the "stuff", as gathered purposefully in conditions suitable for "I" consciousness...in this place.

It seems...those with "I" are at an apex of existence.

• • •

This universe, possessed of physical laws which serve continuance, is not a mind...but it has "mind".

The universe is the stuff we're made of. We have "I" and its self-maintaining parts...why wouldn't the universe have the same...throughout?

Obviously, the universe has "I", in us...but aside from us, if you remove consciousness from "Something", or the entirety, why are there physical laws that allow continuance of the harmonic?

E.

As our singularity banged and inflated, moving into the invisible stuff, atoms eventually appeared, in the cooling. They maintained themselves according to the physical laws formed long before our entry...at least in "possibility".

During entry, when arrived at the time our matter and anti-matter would begin annihilating one another, it was encountering the resting matter of "Something"...which, under this possibility, was perhaps mostly matter...this resulting in the drastic reduction of anti-matter.

• • •

Atoms, as the likely base, maintained and began to be the stuff...mainly gases and suns, while about thirteen billion years passed of growth...awe-inspiring and perhaps fateful, in seeming accord to laws prior to us.

Ionization of matter brought light...say physicists.

Much later, along the way, our sun formed...then Earth... and because Consciousness precedes DNA, Earth allowed hyper-thermophiles, an organism that thrives at water temperatures of two or three hundred degrees Fahrenheit, without sunlight or oxygen, this perhaps the first life here.

Then...entities more familiar to us appeared, and developed.

Eventually, the stuff got housed...sometimes in what we call a skull.

"I" appeared.

F.

Animate things exhibit more than self-maintenance as they move for survival. The tiniest gather by meaningful chance... then improve the result.

Creatures enlarge to the time of having emotion ac-company and be appropriate to going on well.

When arrived at swan-size, as an example, the emotion is an "I-thou" love centered on acceptance and recognition...the swans not needing to think very far beyond instinct.

This is the "set-species" area. There is little "becoming", in the sense that a swan is a swan, a bear a bear, an eagle and eagle. Not to take away from what they can love or be sad about, their knowing what to do usually depends on what passed through the genes, in instinctive behavior, and the uncluttered response to the existence that follows.

...

The stuff of this universe is in all the creatures and maintains itself on a path to body and body-mind.

G.

Man's proper duality has body and body-mind, as one part, and the other part is "I".

The body and body-mind have pure consciousness...and know how to maintain. But they don't have "I".

The "I", or soul, has "mind", in that it shares the "I" of the stuff...and knows it.

H.

We are born with integrity and innocence, resident at the center of "I".

Whatever has come to us through the genes, other than instinctive action, is unconscious to the point of not allowing pre-judged human discrimination before or after birth.

In the womb we are, as to "I", thoughtlessly relative to mother's heartbeat and other sounds, and to physical discomfort and hunger. Then…after birth we are cold because it's cold, and hot because it's hot…and we poop at will, and cry for food. When comfortable we often sleep.

Relative both to this life and the stuff we're made of…

we are tabula Rasa. So, not knowing anything, we don't hate for the wrong reasons.

But…in early years…supplied information enters essentially unfiltered. This entity easily takes in as truth almost all that is put in.

Mist can quickly thicken. Often…in a short time, it's a long way back to the innocent center.

What I'm leading to is called brainwash, or what is put upon a somewhat helpless soul. And I refer to the masses of men.

Institutional certainties, in some places ever-present, can fill to the brim…so there's little room for thought beyond them. Mostly these tell humans what is good or evil…to the point of enchainment.

The journey to a clear place requires much emptying, and most humans are reluctant to let go of what they've been preached to about…one reason being love of the preacher, and/or the conformity he represents.

I.

A friend, Everett Robinson, when reading this part was reminded of the nature of Plato's Cave Analogy, where shadows on a wall are taken as reality.

The brainwashed soul is the caved prisoner...living with information presented as not needing to be opposed. He is seeing what the institution casts upon the wall.

Often man's greatest fear is that his "I" will die when his body does. Preachments, early, effectively remove this fear. Those wanting control...promise immortality, this calling for preached certainties that kill the free will needed to question matters of the soul.

"I"

Part Two

A.

The human "I" is in the head, everywhere. Not only is this focal point beside the right ear and behind the left eye, but everywhere "here"...in here.

This "I am here" would seem different, and more awesome, if the human considerer of it were the size of an atom and standing in his own head. He would say "There seems to an awareness in this entire cosmos. I think the awareness is an "I"...and that it is every-where. How do I explain it?

...

The awareness we feel as being above man...out there, or "just there"...is perhaps everywhere Consciousness, and everywhere "I"

...

"Our Something" has "mind".

Spinning on the head of a pin, our universe is not a mind... but it has "mind".

In my head I have a focal point of awareness..."I".

It seems...that which is the entirety, the cosmos, has a focal point of awareness..."I".

To my logic...it could be that the "I" of the universe is omnipresent, omniscient truth...effective as having once been determined, now being self-maintaining, and allowing freewill to make fate's paths. But the determi-nation was a "becoming of best existence"...this allowing a carpet to be weaved, and all parts to be connected.

B.

As "I" formed..."Something" persisted. Conscious, ness, as love, as harmony, embodied beyond mere "possibility"...in building blocks and physical laws.

...

The stuff of this universe, the "Something" which allows evolution in a place, has allowed the entirety of existence. It's just beyond our understanding.

...

Pure Consciousness, allows "I"-less parts of animated things to know how to maintain and be part of accomm-odation. Atoms know how, molecules do, cells do.

...

Cells in animated entities don't have "I", but they know what to do, being made of the stuff of Pure Con-sciousness.

It is the soul that has "I"...and it is of the same space as the body-mind.

C.

It seems to me that the "I" can separate from the body and body-mind during this life.

When a mountaineer's "I" leaves his body, it is the body-mind, alone, that directs the climb. The climber's "I" becomes an observer...and seems to be a bit away and above.

Whether or not there is an actual departure is impor-tant, but perhaps the main consideration is one of the nature of present control.

One time my "I" departed. I watched my body continue to act, this indicating that my body-mind was in control.

While outside my body, I seemed to be a whole config-uration, and material. I wondered why I couldn't be seen by the others nearby, as I stood above my body. I felt as real, and as awake, then, as you are...reading this.

Most of us feel that we know when we're awake, and that we know we are awake "here", at this point of action. "I think... therefore I am" is an accepted point of existence, but I believe it is also valid to say we know when we are thinking in a conscious state, and not being projected.

In any case...my "I" was an observer of my body and body-mind acting. Perhaps this is the telling truth of it.

PART THREE

You can be between thoughts.

You can be there thoughtless and empty...as before beauty. Or, you can be there even as you think...if clear of being full to the brim with self.

In order to be as a child, empty as to discrimination which twists justice, we need to have appropriate thoughts...or none.

• • •

Inappropriate human discrimination operates far from the center of us...in mist.

Personal allegiance, particularly when felt by youth, usually assures that beliefs will be accepted as certain-ties...closing the door to alternatives.

This can lead to Man interpreting God's will...through centuries...as a power-mongering brainwasher.

• • •

That the soul can close and be cloudy...done by "brainwash", or on one's own, is a danger that surpasses individual happiness...perhaps directing the"oneness" of whole societies.

• • •

What are we between thoughts?

We are the stuff of this universe…loving and compas-sion-ate. We have needed anger, desire, joy, happiness, sadness… without the inappropriate forms of these.

We are near the center of our "I".

Now thinking of a flower…the observed, as is, and our soul, at its center, are in the same clear light.

Away from the center, where we often are, with a bit of cloudiness we can be great, as appropriate servants to others or oneself, according to what love and com-passion can do. But…with much cloudiness we are, as most of you would agree, an inferior lot.

…

Somewhat near to the center, if we can have confi-dence in our worth, and to that extent be full of what we are…yet have an empty place able to let in a value formerly unavailable, or not seen…we can be both full and empty.

Acknowledging the need for an empty place will, in itself, clear some mist and strengthen humility.

Being human we expect to have some cloudiness, but try to rid ourselves of unopposed certainties, and other crap (such as too much of any one thing…even wise sayings).

…

As to crime and punishment, referring to the murder-er, molester, raper, kidnapper, warmonger…those who have victims…we must try to only punish his act, not what he is at his center.

This is a mindset so difficult to attain, man is well-advised to continue to afford justice a blindfold, and let a stranger do punishments.

Ideally, we try to get rid of the murderer's act or acts, as we love what he once was…and, at his hidden center, always will be.

Tragedy is allowed by light in darkness. In a black-ness at the bottom of the worst human action, is a terrible beauty connected to the soul. The pearl in the night…is a child's innocence.

Sure…man has freewill, and is responsible for what he does. And some don't consciously regret their actions,but we need to acknowledge that hidden center. As to the murderer or other bad actor, now referring to his part in the act, there yet exists inside him the empty, innocent place of his mother's womb. Even if he can't feel its presence….we ought to.

• • •

Between thoughts we are loving without the need to consciously discriminate. That's the way each of us enters this life. At that point we hate what hurts…or feel easily in place. "I'm hot…cool me! I'm hungry…feed me!"

We continue to hate what causes discomfort and injury. Hopefully, we get to comfortable places in the right places… that is, not where it seems perfection is, blind certainty is. If we are in a place that can't admit error, or the possibility of it, we're in the wrong place, speaking of inappropriate thought.

• • •

As an adult...between thoughts...after being awed by a sunset or a symphony of sound...why do we so quickly return, and let the "loving only" slip away?

Would you agree that man holds firmly to the comfort of his cloudiness, at least some of it? I mean the lesser paths we settle on, or stumble upon, and get used to, with the foibles thereof.

Perhaps we don't know what we'd be without them, or we enjoy the mere possession, or the "being free to do". Those indignities are ours, personally.

We think it's all right to have a foible or two, trying to keep our sight on the truth of the "all-rightness", in each case. And our foibles sometimes refer outward, as to compassion and justice.

We try to clear away thought's worst parts, and eliminate much of the cloudiness...try to know and feel the uncluttered truth.

Only a few are able to go all the way to the center, stay there, and be enlightened.

All of us are able to be cleared briefly...in any form of selflessness. And yet if we are "loving only"...we can be between thoughts in a way that retains awareness of the self. It's just that we are free of human discrimina-tion. In the sense of being cleared, we are that which is before us.

Then we depart. We leave such as a concert, painting, sunset...or perhaps a sweetheart.

"I"

Part Four

(A)

"Our Something" immediately loved itself, so was sentient, or aware, as it moved, radiating the twoness necessary to continuance.

(by "twoness" I think of what allows such as electro-magnetism and gravity. My sister, Diane, pointed out that, as to the natural order of things, the human will admit the action of "Yin and Yang"...that which gives and that which receives, that which leads and that which follows. And it seems difficult, sometimes, to know one without the other, as to the enabled and the enabler, now considering light and its carrier, the photon...for instance...(speaking as a philosopher).

...

"Our Something" is love, is pure consciousness, occa-sionally shared as "I". Then...the only point for "twoness", or persistence, is harmonic movement for love's sake, for the sake of "Our Something" knowing itself...for existence' sake.

...

If harmony (as perfect imperfection, speaking of what lacks, what needs)…is the basis of reality, wherever in the cosmos "I" can be…"I" is everywhere the same.

Perhaps, in most places the central innocence and integrity of it would be somewhat hidden, as is ours. But there could be places where most entities operate near to the center, with little mist.

...

Between thoughts…now speaking of "I" sharers, here and throughout the cosmos…who are the others? At the center of the "I", and near to it, I am my brother's keeper. The sameness, as to loving thy neighbor as thyself…that is, as to being just and compassionate to one another…is not brought to question.

(once again I feel giddy…this time because I remind myself that "between thoughts" is not often a steady possession, here. I assume it's all right to hope for the best)

...

There are billions of galaxies, and perhaps many universes with their separate singularities. It seems likely that "Our Something" did not originate in a singu-larity.

Also, it could be that many world's exist, in the "other dimension" sense. Almost all humans have believed in two of them, this one and a heavenly realm…either far away, or just above us…or here, where we are.

"I"

(B)

How does one have best contact with the universal "I"?Selflessly. There before greatness, whatever is touched vibrates relative to the greatness.

One is with a Mozart or Tchaikovsky, a Leonardo or Michelangelo…Shakespeare, Tennyson, Keats, Sophocles, Dostoyevsky, Melville.

One sits before a mountain scene, or the Taj Mahal. Or… looks into the eyes of a sweetheart.

A child plays. A loved one relates.

All selfless…the performer and the sharer.

Loving only…without trying to.

Full and empty.

…

There…no questions are needed, sameness being part of the sharer's breath.

…

And at the edge, after one has returned to the mist, what is the height of selfless? Perhaps it is loving all things, the bad as well as the good...even as one fights to be rid of the bad.

• • •

Sometimes the thing before you will tear you from hate... then, in the clearing, you love. That distance traveled will fill you with awe, so that you feel the hand of fate. If you hate... then suddenly love...what is this action other than personal? You feel spoken to...by an awareness "above" you.

From the deepest part of cloudiness, contact can occur suddenly as up not down. There is hate...thenlove. In this occurrence one is emptied, as being "torn" away from". The human "I" is stripped to its edge.

After that...one knows of the cleared path, and accepts that he needs to know of it.

One time, in Nepal, after trekking to twenty thousand feet, then beginning to descend, I became irritable and a bit paranoid...a condition sometimes encountered in those returning from somewhat high places, this due to fatigue and the physiological effects of the whole time.

I acted badly, in a general crankiness, toward my two Nepali friends.

We departed Namche, a village which is at about twelve thousand feet, then within an hour or two, as we walked along the side of the Dudh Kosi, Milk River, they went well ahead of me.

I had entered a cloudy part of my soul.

On the quiet, shaded trail above the sunnied river, I was looking at the ground before me. I...was guessing these men were angry, too...the b....rds.

And I was beginning to enjoy the clouds.

You will agree, surely, that we can slip into discontent, and get drunk on the "old acquaintance" part of it, in the sense of loving it the way some can love alcohol.

The trail went up a bit and began to turn eastward toward Lukla.

There was a small wooden sign announcing the opening of a new trail-side hotel. Ahead, the way was straight and also branched to the left. You could see a building on that branch, and some men working…putting on finishing touches. When I got to where I could have gone that route instead of the usual, shortest way, another sign, halfway to the men, said "This way to Lukla". I knew the main trail went straight, here, but a stranger wouldn't. The b….rds...what trickery!

Yes, cranky! Earlier, a Nepali, thinking I was one of his own, suggested we walk together. As I declined, and he recognized his error, and turned away…I thought "It would have been too bad for you…robbed and beaten!" And at a little hotel by the river, in Phakding, the man running it didn't look at me as he spoke in rela-tion to a question of mine. Why was he avoiding me?

Whew!

Somehow I made it around the bend…where you face Lukla, another village, although you can't yet see it. There are orchards immediately to your right, to the south, beginning a few feet beneath the side of this trail. Rock walls are at the edges of your way, and a forest is to the north where the land is higher.

How beautiful for you! I don't see it! I'm in the cloudi-ness only…and reveling in it.

Suddenly…there is, ahead, that which we all allow. To the left, the wall has been adjusted to accommodate the presence of a Buddhist Shrine, several feet high. Standing next to it is a girl of about years of age. She is smiling. In her hand is a bouquet of flowers.

I looked behind me to see if she attended to some other person. There was no one else.

Beginning to walk toward me, holding forward the flowers, she tripped...dropping them to the ground. Then I could see it was a wreath, fashioned in such a way allowing each stem to pass through a slit in another. Now...with a look of consternation, she kneeled and took the wreath into her hands.

Several pieces were broken. Tossing those aside, she

connected the remaining flowers to each other. Now the wreath was smaller.

The smile reappeared, and she looked at me. At that instant I guessed she was not asking for money.

As I took the gift I was aware of being flushed with emotion, and that I stood smiling in return.

Wanting to respond properly, I didn't know if I could, so I was feeling inadequate. I strained within me to be deserving.

I said "Thank you!"

She bowed, yet smiling, then walked away...on the trail I had just traveled.

I went onward, in a couple dozen steps coming to abend in the trail. There, a well-bundled woman was before me. I couldn't see her face, but her carriage and clothing type suggested she was a middle-age English lady. She said "Excuse me!" in seeking my attention.

That's when I became aware I yet held the little girl's flower wreath at my chest. And, in embarrassment, I lightly pointed out that I was holding it.

Although I could barely see her face in the bundle, the woman smiled, or I should say two lips did, in an easy manner, and said "That's acceptable. Sir...I don't want to bother you, but do you have a wooly cap I can buy?"

"Yes," I said. "At the top of my pack. This morning I took it from the bottom...for no apparent reason."

Putting down my things, I got out the hat and showed it to her. "It's new," I added. "I bought it in Kathmandu, and never did wear it. I..."

"May I buy it?"

"I'll give it to you. I don't need it.

I gave her the hat, retrieved my pack and flowers, and went on my way.

After a dozen steps I was around another bend, and feeling that the little girl and the woman were of a

single unit of action presented to me. I was not thinking that directly...but in my emptiness it was part of the awareness.

Without thought...as I walked...I was wholly of where I was, in awareness, including the time with the girl and the woman...entirely there only...without trying to be. I was loving, as filled by the emptying, without trying to

be. Cleared...I was in a thoughtless love, the first kind.

It seemed...I was where I was meant to be.

With the little girl and the English lady, I was smacked by something more powerful than wind or water...and gentler. In the resulting clear place...I felt free to remain, and be above the usual human condi-tion.

...

At such times it seems that an opened path has you. This is where you become a servant if you stay, in the sense of selfless teaching...although you don't yet know what it is you've learned.

Cleared, I had awareness, as certainty within feeling, that I could stay. Also, I knew I could come back to the usual...and accepted this latter path as it began to happen. Yes, perhaps twenty steps along the way...or it could have been five hundred...I let the clearness go, as my choice.

...

The circumstances of my clearing were dramatic. The emptied condition was of a personal nature, and the change was large...as of a tearing away of uselessness due to twisted human discrimination. I hated...then I loved. And only one of these emotions served the point of existence. Years later, suddenly I accepted the entire unit of action...that is, also what the lady represented of herself. I had viewed the whole as concerning me only, as touched by the child.

As for the lady, the willow tree...the bundle with two lips... it is hard to let go of a chosen lesser path that has become comfortable. As to fate we all have a best path, and we have freewill to follow or not.

Fate brings us to the bridges, including the best ones, directly or indirectly, and we can choose to not cross the best. We can turn away...or, as is usually done...we stay where we are, and the bridge goes away.

Surely, my bad behavior preceding the incident with the little girl was a temporary condition, from which I was torn, by being emptied.

What of the bundle...the willow tree walking? There came a time when I thought, in brief remembrance, how difficult it must be to move from the comfort of two lips in a bundle. What's to take the place of comfort, even the kind far removed from one's best path?

Several years later, in China, if could be that I once again encountered the English Lady. In Shanghai she was yet so bundled I saw nothing but the smile. Then I lightly noted that the smile would come and go, as sadness can touch one's heart.

I thought I saw it. Then she went away. From Shanghai to…I think she mumbled "Chengdu", which is in the west.

As planned…I departed Shanghai for the mountains beyond Chengdu. At the foothills I crossed a meadow- brook and began to climb, which I did until arrived at a great view. I looked back at where I came from. Then…I again saw the English lady, and wondered why. Was my incident in Nepal, with the little girl, inclusive of more than a measuring of my own clarity or lack of it? But I had no interest in the bundle, other than to acknowl-edge minimal value of a fact: I again encountered that lady. Certainly there was no romantic interest. A willow tree? How silly!

The lady and I walked together that day…to a small hotel, the Inn of the Seventh Sorrow, so designated by a friend of mine, a Buddhist priest who has a brother living in these mountains. He said "This is where one decides to let go of a chosen lesser way, or to forever follow it".

I'll shorten the story. Yes…it was difficult for both of us. I… wanted to get to the center of this mystery, or depart from it.

Some of us who accept lesser paths, below what we consider the best available…don't clearly admit to it, speaking for myself and my regrets. As for the lady, she did so admit to it… in her bit of sadness, evidenced in Shanghai.

At the Inn of the Seventh Sorrow one is hiding from a large need. But to open a door and toss something out, what's to come in that's better than comfort?

Of course…the mist deepened before the dawn. Then two I's became one.

Love brings you from a hiding place at the Inn of the Seventh Sorrow.

• • •

Why the lady was hidden is another story. The "hiddenness" perhaps relates to the distance from best happiness, in the lady's heart. If she regretted being a willow tree, but accepted it, the "other story" is sad.

At the Inn of the Seventh Sorrow, found happiness adds some point to prior sadness, in the studying. Now, in happiness, maybe the sadness brought her to me.

• • •

Concerning fate, I believe the purpose of "Our Some-thing" is to continue to exist, at its best. In that sense it is "becoming"...as to self-maintenance and onward, within best harmony.

If reality is a tapestry, it is a "becoming" one. I think all is not written in the weave, but being written.

In the way humans are "becoming" throughout their existence...fate brings you to the bridge, but in your "becomingness" you have freewill. When you get there, even if you think crossing it is your best path, you are free to not cross it.

And so...we are humans and not robots.

As for the tale...

The lady hid an inner song.

("care" gets beaten and hides its head)

"I"

Part Five

Of "reality"...we have some acquaintance with this universe. It's our area of study. Our imagination can go farther...so, for this universe, as "inclusive of", I see the following as being true.

Sentience, as universal "I" and its parts, resides in all things.

All things are harmonic and rhythmic in essence. Thus, at the base we have atoms, strings, or other rhythmic unit, perhaps operating in salvo...bang, bang, bang...maybe all at the same time.

Chaos and aberrance, possessed of harmonic parts, at least in promise, can occur only temporarily, as not the natural order of things, being pointless.

That there is "Our Something" rather than "Nothing" says self-maintenance, in all its aspects was, and is, desired.

The purpose of the entirety of "Our Something" seems to be "becoming" in a way that necessarily moves by its laws... but is determiner of the laws and, so, rests above them.

• • •

In the animated, pure consciousness is present in a way above maintenance-only, to the point of attempting to improve upon, or at least adapt to supposed good.

Whatever worlds exist...the parts of each will have pure consciousness, but gathered parts of it may or may not be capable of having "I".

• • •

In our place...sentience, as "I" and its parts, enables us not only to exist but possibly to do it well.

Here, the consciousness of the animated touches the material world...and it does so in a way beyond our knowledge.

I say "I'll move my right arm"...and if I choose to move it, it moves. My "I" touches the parts of "I".

Perhaps consciousness, as thought, radiates to effect, here, just as it does at non-zero degrees.

• • •

Is Consciousness the Ground of Being?

If it is...love of existence is.

Consciousness is "Here" everywhere, only if it wants to be.

• • •

The world is perhaps more connected, one part to another, than we thought.

"I"

Part Six

Referring to this dimension, perhaps a Ground of Being is possible if division of matter downward ends at consciousness... which lacks, which needs, which has purpose within imperfection. By another name...I'm referring to love.

...

(Aside...

Because "Our Something" is a throughoutness...it seems to be as a tapestry, continuous, connected.

In our universe time is felt to be one thing following another, related to the speed of light. In all reality, which includes us, time and distance is the change occurring in the continuation of consciousness, as it goes forward the "same as" and "different from" within each unit of itself.

I'll speak about this later)

...

The clear center of "I" can be buried, but it can't be beyond reach...because freewill survives with "I".There is fate and "brainwash" only to the bridge. If we had to cross, we wouldn't be human, and "becoming".

...

Man can feel heightened contact with the universal "I" in special circumstances...and most easily when rendered empty, free of human discrimination.

One can be drowning and loving only...that is, loving only life...good and bad.

One can be thoughtless before beauty.

One can romance a loved one...be free there, as to lost and found, in the emptying...and return to the usual a clearer soul.

...

"I" is an apex, or summit, of awareness.

It ought to be assumed that, throughout "reality", "I" is a crowning achievement.

Every suitable, long-lived place will have "I".

"I"

Afterword (A)

During much of man's history, the use of consciousness has been dictated by major institutions.

In action..."I am better than you" has reverberated through the centuries to us. Some institutions, notably "of church", and now I repeat the word "some", have carried that tenor silently, calling it "love thy neighbor"...as entire continents fell to domination.

...

We are made of the stuff of the universe...the laws of which are harmonic, or we wouldn't exist.

At our centers we are equal...and we can see the truth of this in the study of its basis.

...

Among the masses of humans this center has been often buried...as billions have been spit on, cast aside, beheaded, blown up, stoned, burned, stripped of property.

Some institutions have floodlighted the dark parts of mind...euthanizing the fear of mortality...this assuring the power of dominators.

How can we rid the earth of this?

Also...as we emerge from the dark ages, how do we avoid moving into the new darkness...now threatening?

It is the tendency to study, scientifically, the human body and mind as though man is machine, devoid of freewill, with actions pre-determined by conditions of genes.

Institutions have controlled choices, for centuries, turning many into sheep. "Scientific" study can do the same.

"I"

Afterword (B)

As we emerge from the darkness...we must study, as physicist and philosopher, how the body and soul of man relates to the awareness beyond himself.

• • •

Why do we have "I"? If we are made of the stuff of the stars, what is throughout existence which allows "I"?

• • •

There seems to be a large awareness beyond and "above" man...as part of man and all existence.

Perhaps we ought to relate its nature to what we see of ourselves and what is around us and to every new thing seen under the sun(s). In the wilderness rests a sameness in all things. There is an equality that refers to the definition of the awareness...which is love of exist-ence, and of what exists.

• • •

Man has the freewill to see reality or to close his clear-minded empty place. If he chooses to live far from the center of himself...in the clouds of "brainwash"...he is free to do so. But when he does so, he is, effectively, a robot.

• • •

Man's make-up shares the laws allowing the cosmos to be.

For the consideration of science, if the cosmos is not a machine...neither is man.

Freewill defines man if Consciousness is the central determinant of physical laws...ones that allow the existence of all things, and the continuation of them.

As we emerge from a dark time...we ought to think about this.

Afterword (C)

If we think there is no awareness beyond us...one relative to all things...we can enter the age of robotism, here...and become slaves to central power mongers much worse than the ones we now have.

AWARENESS
AS THE GROUND OF BEING

Alfred John Dalrymple

AWARENESS AS...

Contents

(A)

PART 3

All reality is "becoming". You can't find the Buddha "arrived".

PART 4

Awareness/Consciousness...is it the Ground? There is self-maintenance according to the laws, because all things are the laws. They have pure consciousness, which is a non-cognitive sharing. Man has both pure Consciousness and "I".At their innocent center,

AWARENESS AS...

Contents

(B)

PART 8

Because all existents "effect", they have at least non-zero mass...said the philosopher.

PART 9

The path between man's innocent center and the "I" of existence, or God, is almost always obstructed, a little or a lot.

PART 10

Action-at-a-distance, when not relating to the speed of light, pertains to Awareness/Consciousness.

AWARENESS AS…

Contents

(C)

A seer sees "between thoughts".

PART 14

We are imperfect. We wouldn't exist if we had no lack, no need. Consciousness is the stuff of this universe, and of all existence. Man's "best existence" is "loving only". But…few humans have been able to love all things. Most of us

AWARENESS AS...

Contents

(D)

focus on loving the good.
Fate and freewill.
Both are true.

Here and everywhere, I am "this man" and no other...while trekking the rocky road.

AWARENESS AS THEGROUND OF BEING

Alfred John Dalrymple

What I know is stored in
Zithers...pluck, pluck, pluck

Imagination is greater than knowledge.
Albert Einstein

Part 1

With sincere logic we imagine beyond knowledge. But if our knowledge seems bizarre, even though accepted, we hesitate to dream.

We can be robotic in hesitation, if certainties hiding in "acceptance", block new thought.

One of our certainties is that nothing can exceed the speed of light. We refer to this limit because it accords with our understanding of time. It seems that one thing follows another in this universe.

This is the shoulder we stand on to see if message is sent between separated things.

• • •

In consideration of the nature of reality, what do youmake of the accepted knowledge called entanglement?

They say bits of matter or light can, even worlds apart, act as one. Think of two subatomic particles that are somehow connected or "entangled" with one another, so that when something happens to a particle, the same thing happens to the other, even if greatly distanced. This concept violates the theory of Special Relativity, which holds that communication between two places can't occur faster than the speed of light.

• • •

With entanglement…it is obvious that responses between bits might not relate to the speed of light.

Considering bits apart by vast distance, immediately co-responding to stimulus applied to one…because we have joined the speed of light to the possibility of having message, we believe, usually without question, there was no message sent.

• • •

It can be said that one thing following another is "time" in our universe…but, also, it needs to be said that if our universe is part of a larger reality, separated bits might have message sent in a way different from what we now understand.

What can carry message immediately, even over vast distance…and be relative to this universe and also to a larger reality?

The following seems likely:

If existence didn't begin in this universe…"time", accepted as one thing following another, limited by the speed of light, can also be defined by laws serving a different means of communication.

• • •

If my Zither stretched to “all”...
this finger at light’s command,
one pluck...

AWARENESS...

Part 2

"Our Something", once existing, continued. It must have wanted to", said the philosopher.

The proper question:

Why is there "Our Something" rather than "Nothing"? When dealing with existence and the either/or of the above question...man has a problem with "Nothing". It isn't there... or is it?

• • •

There is "Our Something"...and the "Something" it came from.

Of the part of "Our Something" we know a bit about, this universe, they say it began in a singularity...which banged, inflated, expanded. Why did it do that? And why was it there? (exploring "possibility" as the first ethereality)

It seems...after cooling and ionization...if the speed of light didn't limit dissemination of laws within this universe, even considering repeat performance, the nature of a larger reality explains the presence, and the "possibility", of laws.

• • •

Considering this universe and a reality larger than it, any Ground of Being is going onward to whatever end it contains or grows toward. Even if the end is a wholeness, the journey is one of imperfection. There is lack...and, so...need was at the beginning. If I'm here, I need to be there, or journey, itself, would not exist.

At the beginning, if "Our Something" wanted to con-tinue, it necessarily, and immediately, disseminated laws allowing self-maintenance and growth, within, and as, a medium of its own nature.

Am I describing the Ground...Awareness/Conscious-ness?

• • •

Consciousness immediately was of its own being, meaning there was something love, which was Con-sciousness aware of existing and, so, aware of itself.

Consciousness...immediated laws allowing continu-ance, allowing "becoming" onward.

(Is energy nearly free-flow...at non-zero degrees?)

• • •

The proper question, once again:

Man's knowledge makes it valid to wonder how "Our Something", once existing, continued to exist.

"Our Something" persisted when laws allowing existence were everywhere immediate...this in order to be, immediately, the Entirety.

• • •

As in Entanglement, where separated bits of matter can, at any distance, act as one...we should consider that the Ground of Being ought not to be identified as needing the acceptance of our present view of "time".

Perhaps only within a containment, such as this universe, do we need to refer to the speed of light...thus to time and distance as we know it.

At the Ground...now as to the Entirety...perhaps there is no "A" then "B". "A" is "B"...except that within this carpeted sameness is "difference", no existent being "arrived". The Entirety is necessarily imperfect "onward".

The Entirety of "Our Something" is "becoming".

...

As to that entirety...we "become" in a continuance of both sameness and difference. The "difference" at the end is contained in the beginning. We are connected as on a carpet of Consciousness where, although "become" is time's porter, being relative to the carpet identifies time, or placement "here and there", as immediate.

...

1.

The condition of the Ground.

The Entirety exists because when "the nature of A and B, as to why they are energy or matter or both" is part of the answer to why we exist, "A and B" doesn't only relate to the speed of light but, also, to the condi-tion of Consciousness.

2.

Are the parts of reality separable?

Considering two separated bits acting as one, either there is message sent between them or there is no message sent.

If there is message sent, how could this occur faster than the speed of light...and not interfere with proper time, one thing following another, within a contain-ment?

What is faster than the speed of light, and a necessity to proper existence? Or...what is the nature of the con-sciousness you use to wonder about this?

3.

"Our Something" continued...and continues.

If two bits of matter or light, separated by vast distance contact in a way considered to be "message", validity applies when sincerely trying to answer the question about how "Our Something", once existing, continued to exist.

The Ground...Awareness/Consciousness...dissemi-nated laws allowing continuance in a "here and there", timeless and distanceless way.

Considering vastly separated bits, if message is im-me-diate, what is the distance between them? The question de-mands moving our consciousness beyond this universe. As to entire existence, it seems the distance is zero.

But we need to remember that "difference" exists, and this allows cause and effect...thus there is a sepa-rable nature to reality.

...

Consciousness allows imperfection...onward. I think this is so, because consciousness "Entired" immedia-tely...or there would be no laws.

"Becoming" is allowed by the allowed laws, in a sometimes seemingly chaotic, and always imperfect existence.

4.

The materiality of consciousness.

As to the so-called ethereal and material, we observe the apparent difference...but the sameness is now beyond our sight. Helpfully, perhaps, we ought to think about how one creates and/or touches the other.

How do you move your arm by deciding to do so? What is it that precedes known stimulus, for instance the electrical or chemical? Whatever we have per-ceived as being the initial action involved, how was it touched by the thought?

Has man acknowledged the contact?

• • •

It seems...all existents have movement, have at least non-zero mass, have materiality...in that they effect other parts of reality.

I believe...materiality is an aspect of what conscious-ness dictates...and has dictated...in laws.

PART 3

The Ground "becoming"

Awareness/Consciousness loved existing…which concept includes "continuance".

Consciousness effected laws for necessary "twoness"… this in the way of "other than oneness", so "equal to but different from".

Consciousness knew that continuance would happen at allowance of laws…and knew that "becoming" is necessary to continuance…thus perfection wouldn't kill growth within ethereal materiality.

"If you think you found the Buddha…kill him!" said a friend of mine, referring to the Buddhist admonition. I assume they mean that when the Buddha sat under the tree he became "one" with a "becoming" thing…so, also, the Buddha is "becoming". That's why you can't find him "arrived".

That's why you can't find yourself arrived, but only "becoming".

PART 4

Condition of non-zero mass

A.

Sentience...as Awareness/Consciousness...resides in all things.

Self-maintenance is throughout.

All entities and their parts and their parts, inclusive of all parts, including space and its reason for being "not nothing", have self-maintenance according to the allotted laws of the Ground.

B.

"I" embodiment.

Perhaps Consciousness embodied as Entirety "I" in a way similar to the "I" embodiment in the human head, "Here" everywhere.

...

All existents and their parts share "I" as pure con-sciousness...non-cognitively. Atoms, electrons, quarks, have pure consciousness. They have self-maintenance, and serve "best existence" according to the laws.

• • •

Some existents, notably man, share "I" cognitively. They know they have "I" to the extent of being able to define it.

• • •

In my head there is "I"...as an entirety. No parts have "I". My "I" is not any part, but is all of my soul.

Also...man has always felt a "Presence"..."just there" or "above", of consciousness which seems barely out of reach of the understanding, and which seems to be an embodiment of the "I" of it all...or God.

C.

Pure consciousness.

Some entities can directly share "I"...but all existing things and their parts, inclusive of the parts of a man's brain, have pure consciousness.

Pure consciousness is a non-cognitive sharing. The parts of entities that have pure consciousness, have non-cognitive knowledge of Awareness/Consciousness.

Consciousness, as to its allotted laws, is "best exist-ence". The harmonic is not only necessary to existence but is desired, non-cognitively.

Perhaps this is a form of non-cognitive freewill.

Parts are free to do the action of the laws...in that "they are the laws". And no self-applied discrimination will stop them from being the laws. They have no self, other than being "of" the laws of pure consciousness.

Entities with pure consciousness only, maintain, thus "become", without interference from cognitive discrimi-nation. They are of the laws...so have no other thing to do but be "best existence". They love "best existence", non-cognitively.

D.

Sharers of "I"

Some entities, namely man, have both pure con-sciousness and "I".

The "I", or soul, is enabled by the stuff of Entire "I", as gathered in the body-mind...wherein there has been established proper conditions for it. The "stuff" is that of the universe, thus of entire existence.

The "I" is housed in the stuff of the body-mind but, once allowed, the "I", or soul, is separate in its entirety.

...

The "I" is separate from its place, and within its contact with Entirety "I", as sharer, has its own freewill.

Although Entirety "I" can be called God, it ought to be accepted that man is responsible for his choices. Without separateness, as to freewill,man would be machine.

...

Awareness/Consciousness...assures freewill. Consciousness, free to be of "this" man and not of that one, can never be machine. Man has "I" consciousness and, so, is free to choose what he thinks is "best" but, also, what he thinks is "worst" existence.

If man's pathway, that which connects his center to Entirety "I"...is clear, is not cluttered by inappropriate thought, he will do what is there to do in a loving and just way. So...although alone...he is, at that time, brightly connected to God.

E.

The self...and equality.

When man's path is fully clear, this can be called Enlightenment. And, although a few seem to have acted above that power, and deserve different characteriza-tion, because man has freewill, even at the clearest time he is "this" man and no other.

In the matter of administering proper justice, if there is no self, there is no other...and, so, we are equal. But the self I, or "this" man, must filter to its essential nature is the human mentality. If I can sweep away false thinking, I will be just...in my love of what is right.

The human self can always be closer to being "the same as" what is at the other end of the path. Although separate, I can improve my sense of equality.

Cleared, to a reasonable extent, and yet alone in my choic-es...I am Frank not Paul, but I will treat Paul as

though we are the same in the wilderness.

PART 5

Some concerns to Evolution

As man's "I" touches parts that have pure conscious-ness only, this as to the body and body-mind, what changes are effected?

Today…I think I'll walk on two legs. So…today, I do that. Of course, my legs are the same length as my arms…also, they're crooked.

Tomorrow, will my legs yet be crooked and short? How about next year? So, what's the point of trying to stand? Maybe I'll just get a ladder.

• • •

Does the "I" with its perceived "best existence"…that of standing erect…touch the non-cognitive freewill of those parts possessed only of pure consciousness?

The cells of the body know what to do to self-maintain… they have potassium in proper amount and get rid of excess sodium…without our "I" directing them to do so. But…do they know indirectly the per-ceived wish of the "I? Consciously and purposely I direct "whole part" movement of this body when I stand erect and reach for a fig, or look over the high grass of the plain.

I don't consciously know what my cells are doing to maintain. I sometimes think I do, because of some knowledge...but I don't. And I think that my cells don't know what I wish for, but...

Over the ages someone has occasionally said "We should be careful what we wish for!"

It was good for us to stand erect. With consciousness we make a change of usage of what we have, by using it differently...then the cells of the body change what we have...or a gene will switch on or off.

But...our "I" is in the habit of sometimes making wrong choices. The human mentality can get in the way of the truth of things, concerning what is best. Our path to Entirety "I", or God, as "best existence", can be buried, a little or a lot.

Is it good for man to get bigger, now? There will soon be nine billion of us reaching for a fig.

Can those things possessed solely of pure conscious-ness block the wishes of the human mentality? Not over the short run, in most cases. If you feed your face all day, the body will get bigger. If you choose to not eat, your body will get smaller. But...over the long run, the body can effect change in self-defense of a threat to "best existence". Maybe it will find ways to have the human brain's capacity shrink to the point of not knowing the result of an invention. Then...man can fall merrily.

PART 6

Body-mind...as able to act alone

The "I", or soul, is separate from the body and body-mind. And it can be noted that many believe the "I" can depart from the body-mind during this life. They think it can be done purposely, but that the occurrence is usually accidental, in most, and rare.

...

While reading a book aloud to a class of fellow students, if a man's "I" departs his body-mind and goes to a position just above...and, there, observes his body reading...who is reading?

If "I" am "up here", and not consciously in my body, "I" am not in my body.

...

I don't mean to suggest that the body-mind, without the presence of "I", is robotic. Even if the body-mind doesn't possess freewill of its own the way the soul does, it is capable of self-maintenance, and can improve itself due to having sentience attuned to "best existence".

If a robot enters an extremely cold place, without having been programmed to react to cold, it will likely be damaged.

The body-mind, capable of being in charge of the body, has the innate need, served by memory and a form of creativity, to put on a coat in order to provide warmth, and would direct action to servethe "best existence" of the body as a whole.

• • •

The body-mind, possessed of pure consciousness, acts according to best, or at least adequate action in relation to its area of engagement. If a human is reading a book, aloud, or driving a car, and his "I" departs, of its own volition...the body-mind will continue to read or drive.

It is sometimes reported that a mountaineer, facing a near-death situation, perhaps clinging to a rock, will leave his body and be just above it, as his "I" observes the situation safely, and his body and body-mind contin-ue the climb.

PART 7

Awareness...and "this" man

DNA has non-cognitive knowledge of "best existence" for this unique thing it is...and it shares the freewill of pure consciousness. It doesn't share the freewill of any "I", but it does have allowance to carry forward the physical and mental uniqueness.

While the "I" is departed, the body and body-mind direct the body to act, and it will act in a way unique to what DNA has provided.

At the bridge, if my "I" has departed...my body-mind, as partially established and captained by genes, will serve "best existence", or maybe second best or third best...by non-cognitive freewill.

When my "I" is present it directs consciously, by its own cognitive freewill. At the bridge my "I" takes precedence, as to choices, over the body, body-mind, and Entirety "I", or God.

If my "I" is at the bridge, I am there as Paul, and I can choose to do "worst existence", which could mean doing nothing... even in the face of knowing action is called for. Thus I am human and not a robot.

...

Those things with pure consciousness only, including DNA and its parts, can't choose "worst existence". If a cell self-destructs it must do so to serve perceived "best existence"...otherwise, it didn't make the choice except as being acted upon, and thus responded as best it could to stimulus it couldn't properly assess.

• • •

Man can know about existence and define it. Also, he can make a joke of it or a game. He can choose to be wrong so others will stop hating his intelligence, or to have integrity to the point of ridicule or death.

PART 8

Movement in the mist

Speaking as a philosopher...

It seems all existents have movement within...have mass, at least non-zero...and could be said to material.

In a field of so-called massless bosons...spin can be imparted to the passerby, or the merely encountered, increasing their mass to at least non-zero. This suggests that bosons also have at least non-zero mass.

...

Bosons have freewill of pure consciousness, and impart spin according to the laws of existence.

Bosons effect reality by touching it.

It seems that Dark Matter and Dark Energy, whatever they are, effect reality in a way that, as described, also uses the word "impart"...and could be said to touch reality.

Your mirror image effects by touching your opinion, which reaches for a hairbrush.

...

As a philosopher...I am trekking the rocky road of consciousness. It seems to me that all existents have materiality, and have been allowed by consciousness, even if the first so-called ethereality was "possibility".

Consciousness has movement, has non-zero mass, has materiality. This is because consciousness, even as "possibility", is necessary to existence.

• • •

For all practical purposes, the laws have been allowed that when I stub my toe it hurts.

PART 9

The nature of contact…as to the pathway

Man contacts deeply with Entirety "I" when…

1.
The pathway is fully clear, permanently…this referring to the enlightened.

2.
The pathway is very clear, permanently…this referring to the sage.

3.
The pathway is temporarily clear, while being "of selfless-ness"…this referring to most of us.

…

Perhaps the easiest way to visualize what I mean by self-less, is to ask: "What am I between thoughts?"

I move this question beyond childhood, where loving life can mean they're at the pond chasing frogs, the point being that innocent ones love the beautiful not the ugly, and go astray by accident…within innocent purpose…and rarely very far.

The adult human has long acquaintance with universal standards of the "beautiful", as it matches his appreciation… then, with deeply buried innocence, can happily describe how he gutted it.

• • •

What am I between thoughts?

If I am the stuff of the stars and of all existence, I'm most purely that when free of the worst parts of the human mentality.

The freer I am the more easily I love justice and kindness and harmony…even while admitting the necessity of imperfection. I…love life. And, so…I am connected, in being nearly what Entirety "I" is.

If I am between thoughts, and before beauty, what else can I do but love life? I'll be as a child again, there at the innocent center.

• • •

How to be emptied? One doesn't need to work hard to find that opening. Most of us encounter rare moments every day… with a child or other loved one, sometimes rapturously so, as with a sweetheart, or with nature or a work of art.

Why point this out? Because the clearing usually doesn't last long and, sometimes, when man comes away from these moments the truths immediately are detoured into reinforcement of earlier prejudices. He can be farther away, than earlier, from the clear place.

Referring now to the first half of the twentieth centu-ry, it seems that the two most cultured nations were often making war. Germany and Japan. Why? And…is there a mindset we can attach to the cause?

Being in contact with the Ground's actuality of why and how we exist, is to be with love and equality. At our centers we are all the same in the wilderness.

But…unfortunately…the truths in the contact, as we get cleared, are easily twisted and hidden as we return to the usual. We can readjust to a filled pond. For many in the two nations cited, the "fill" was a sense of superi-ority. A return, immediately, to dangerous unquestioned certainty usually means it's got added strength.

"I am better than you!" is one of the steadiest and most evil of man's tenors. And now, of course, I consider what one thinks they deserve of love and justice, in relation to what others deserve.

•••

Be as a child again…but it isn't easy to do that. Children are usually "between thoughts", or innocent, even while thinking. Adults are often busy trying to fill the self and others to the brim.

In the sense of that "filling", man sometimes thinks too much and poorly. The truths of reality remain far away, including far within, effectively out of reach. And one wonders what good it does to have a few cultured moments.

What if an institution, here referring to a nation and most of its members, have a mindset celebrating supe-riority rather than equality? And the superiority seems to be mostly "theirs only".

That would be a long story...inappropriate to an essay not primarily dealing with fear and brainwash, or with the pleasure humans can feel when thinking they're better than others.

Speaking of those who have some freedom to self-direct, and have a strong grip on how to properly serve all men, in the development of that self...once back to the usual, what is it that blocks the path?

What are the worst parts of the human mentality?

After acknowledging the value of anger, desire, joy, happiness, sadness, the appropriate forms of these being necessary to our best existence...we need to know that the inappropriate forms can be guilty of "filling to the brim".

One can "dwell upon". And in the institution, the hurtful certainties can rest on what the ancients called

"overweening pride"...this to the point of hurting others.

But...now, I'll address, briefly, how most of us misuse the usual emotions.

As for me...if someone for no good reason, it seems, steps on my foot, I get angry and push him away. Then I let my anger fade in relation to the distance between me and the nitwit. But if I dwell on the incident for a length of time beyond the value of helpful result, and just do it to retain anger, obviously I'm not serving the best existence of me and my society.

We know about incorrect use of desire and happiness. Awry, they have centered many a bad end. As for sadness, dwelling on it shows a hurtful tie between the self and "letting go".

We study history, both of the institution and the personal... then we rid ourselves of "self-celebrating judgment" about it. We remember what we loved of it in a loving way...most fruitfully by letting it be itself, as not needing to be judged. The parts of it we see as hurtful we try to hate hatelessly...in the sense of loving life, the bad as well as the good, even as we move forward trying be rid of the bad.

With good ethics we want to do what's right, not according to "I am better than you!" but by a feeling of love and equality. Left alone in this freedom we can feel safer in errors of human judgment. Even being as a child again one can do wrong, in a "between thoughts" way, and the caring person can catch a frog down by the pond. Of course, as one matures, letting the frog go can be called compassion.

PART 10

Beginning to remember about the "stuff"

Things separated by vast distance, when seeming to act as one, have definite states or characteristics of their own.

I believe action-at-a-distance depends on message. If message is immediately available for separated bits of matter or light...according to the condition of Conscous-ness...reality retains it separable nature.

• • •

I think our world is both real, existing without our observation of it, and separable, in that "connection" is caused.

PART 11

Cognitive and non-cognitive freewill

Because our universe exists and was once what they call a singularity, its seems to me that there were laws allowing it to enter and move. There continued harmony within imperfection... even through the seeming lawless-ness. Sure...probably laws were present before we popped in.

...

If conditions allow, with pure consciousness all things maintain and are part of growth through non-cognitive freewill.

Awareness/Consciousness precedes DNA. But the parts of DNA have pure consciousness, as self-mainte-nance and as part of that which "becomes"...and they are possessed of non-cognitive purposeful maintaining and becoming. They don't initiate by self-emanated choice or freewill; they seek continuance through becoming "best existence", as part of the "I" of reality. Choices of DNA are made through the sensations allowed by Entirety "I".

Sentience, as consciousness, resides in all things...as they maintain.

...

If the "I" departs the head, temporarily, the parts of the body-mind allow this entity to continue acting, without "I", and only with non-cognitive freewill.

Only those possessed of "I", the entirety of the sharer as one unit...have cognitive freewill. And each sharer is unique. When I come to a bridge, I cross or not according to what I choose within my "I" conscious-ness, not only as a man but as "this" man.

...

As to entanglement of many, such as demonstrated by a group of birds or fish, there seems to be action through message, somewhere beyond the need to think about it. But even at that, if they act as one being, perhaps there can only be message sent by the condition of shared consciousness.

Delivered at the speed of light, a purposeful intention would precede message. But...within the condition of Awareness/Consciousness, there only needs to be ema-nated consciousness that a "turning to the right" is now occurring.

...

If the "I" of "Our Something" is imperfect, this allows it to be "becoming" onward. Our "love"...is defined as needing "more than". Does this mean that even if man is entirely open, as a full sharer...and as a consciousness "different from"...he will always have freewill in the "becoming"? As long as existence is...at the bridge I will always be alone in my choices, as separate from God, but also as being Frank not Bob...thus separate from other men.

...

Man's possession of freewill gives him the luxury of being wrong, sometimes...in a necessarily imperfect world.

Our new pathways might lead into mist...but how else does man advance? He does so not only through adversity, as Joseph Conrad said, but also by necessarily being wrong, occasionally.

PART 12

Necessary imperfection

A.

Can a machine, human-like in action, be conscious?

The human mind has soul, or "I", as it stands, and then information enters. That information doesn't create "I", but merely helps characterize the whole.

As for a robot, information put in doesn't have "I", or become it, nor does any which is gathered.

Words entering a robot, as stimulus to action, come from an entity possessed of "I", but when the words are apart from that "I", and contact only that which doesn't have "I", they are stimulus only, and constitute "formula", which travel relative to the speed of light. They are limited to areas far below the possibilities resident in "I".

A robot can act, seemingly, on its own. Formula is registered and connect to places of the electric, or chemical, to spur movement alone or in conjunction with other entered directives. But...perhaps the reason "I" can never develop in a robot is because the machine's action, as to conscious being, can't travel faster than the speed of light. The "thinking" of machine can be "here" then "there" according to man's allowance, but not according to self-emanated con-sciousness, which requires possession of "here" and "there" of consciousness.

...

Some...who think freewill can be programmed into a machine, are quick to believe that when we come to a bridge we turn left or right according to what our genes dictate.

...

Before having a word about DNA...it needs to be considered that, although self-maintenance is a result of the laws, all units that gather to make a larger unit are acting as pure consciousness. At least non-cognitive self-directed purpose is involved. When we consider a computer, or a human-like robot, the gathered entity was put together by man or a machine made by man...not by its own parts as pathed, and acting in conjunction with, the "stuff" of consciousness.

...

Considering stem-cell research, perhaps a boon to curing illness or other injury, it is forecast that they will

soon have the ability to make a human brain.

Can that brain have "I" consciousness?

Several decades along...as you walk on a street, you might wonder if the approaching figure has pure con-sciousness only, or also has "I".

If we are visited by creatures from another planet, will we encounter one fully robotic, or perhaps one having pure consciousness only? Would it be more acceptable to meet one made from a stem cell?

The question I move forward is: will any of them have compassion?

When "I" isn't lineaged...a creature will do what it is supposed to do, relate to "the program", or to "best existence" within its available or appointed area of action.

• • •

If tragedy is allowed by light in the darkness...the hope at the end of the tunnel...is it only sharers of "I" that have "that" darkness and, with it, true compassion and love?

B.

Being conscious of one's imperfection

A machine can't be necessarily imperfect, in action as a whole, in the sense of imperfection being within harmonic onward movement. A man is necessarily imperfect in order to exist. Humans share "I" with Entirety "I", which has to be imperfect to continue.

All reality exists because of imperfection, which means lack is part of growth; but not all existents have self-emanated consciousness of lack.

The seeking soul has continual missingness, and the "I" consciousness can try to identify the purpose or the meaning of what's missing in relation to happiness. Those with "I" try to know, in the sense of "caring", where they're going.

• • •

Imperfection is necessary to the beauty of light in the darkness. With the darkness comes love and hope. The darkness, itself, is "becoming" toward the light. And as a place for it.

By loving...we "become" as we seek harmony in imperfection.

Without imperfection there would be no need for consciousness.

C.

It's likely that more passes through the genes than we know about. We accept that images move onward...in instinct. Perhaps, also, words and entire scenes pass.

If man is made via stem cells, what of "ancestral knowledge"? If a man arises from human DNA, and this of one individual, will it pass forward more than physical characteristics and the pure consciousness of the body-mind?

Can a cell pass forward images, or words, asleep to the man's consciousness, including those which were passed for generations?

If a man is made from a single cell, perhaps it would develop enough gathered "stuff" to have "I" conscious-ness, but can it carry forward the stored truths of the sperm and egg?

•••

Considering the making of a human brain...what are the benefits? Perhaps if my child, dying on a hospital bed, could be given a new brain made from his own stem cell...?...this somewhere in the future.

What injury can come to man because of this?

As man makes a man...will he only pretend to care about the consequences? If he truly cares, how patient will he be?

Obviously...George Orwell (1984) and Aldous Huxley (Brave New World), in particular, were concerned about the "caring".

When I imagine the man-made man before me, I see him as being taller, stronger, handsomer, more sociable, and smarter than I am. So…who needs me?

Speaking for all eight billion of us.

PART 13

Fate and freewill

A.

Sometimes man arrives at an event at which he/she has no choice of action. For instance...a person walking on a city street is struck by a falling piano. Also, it could be that there is no apparent connection between the piano, the dropper of it, and the head that's hit. For all practical purposes it was unavoidable and accident-tal.

Occasionally...one has a bad feeling about walking the street, so chooses to stay home. He/she decides to face "best existence" by avoiding what could be a harmful rock in the river, or street, of life.

We accept that each life will contain parts within the wide range of "chance". There are bridges you will come to, and some will have action beyond available choice, like that of the falling piano. At other times after having a bad feeling about the tenor of a time or happening, you can choose to stay home.

A consideration of "fate"...now enters.

The masses of men can get a useable view of fate by reading Sophocles' great drama "Oedipus Rex", in which a seer sees the future in a way that is seeing what can't be changed. What is seen is that he, Oedipus, will murder his father and marry his mother. As the drama unfolds, it is clear that nothing will stop the "foreseen" from occurring. Not only do all the char-acters connect but so, also, do the events. As they occur, Oedipus is unsuspecting, but it is clear to the reader that his drama's connections are so powerful, fate could be proved by them.

• • •

Some thinkers over the centuries have felt that "all is written"...and would argue that even when choice seems available, you don't really have a choice.

• • •

When choice is available, to say "all is written" seems to identify man as being robotic...without freewill.

Sure...I believe the bridges you arrive at that are not "falling Pianos", you cross or not according to your choice...not DNA's, or God's.

Also, if fate is described as aftermath to a whole, now meaning when a person is gone, whatever happened, in report, couldn't have happened any other way than the way it happened...is a copout, I believe, to robotism. A whole had "this way and that" within it...a best path or not...a bridge crossed or not...in an accordance to your choices, as an individual consciousness acting alone and separate, in an imperfect, "becoming" reality.

• • •

To have a future event seen for you...seems to deny freewill. When an event seemingly foreseen comes to pass, some might feel they are a machine in a written reality.

I am not a machine. When I arrive at a bridge, I cross or not according to my conscious choice. I stand there, as this "I"... alone.

• • •

As for tomorrow's "foreseen" battle, and whether or not the combatants have a choice to attend or not, is matter for debate.But some would say whatever the seer sees...will happen...that there is no choice. If she sees you wounded there, you will attend and be wounded.

I agree...with qualifications.

I believe that the seer, between thoughts, can be of a future event...but what she sees will have been deter-mined to that point in time by the choices of freewill.

• • •

Consciousness is immediate as to the carpet, and yet it is "becoming".

Consciousness is the Ground of Being, but our use of the word "time" must be applied both to this universe and to the Entirety beyond it.

I think there is both freewill...and...fate...in that even though a future event can be seen, it will have been allowed in a "becoming" way due to freewill.

I am not a machine...I am human.

B.

Fate about to be viewed in ways different from the usual. Entanglement is the condition wherein bits of matter act as one bit, even if separated by vast distance. This negates the speed of light within "message". So...when you kick one bit and the other instantly responds, it is believed by many that no message was sent, which denies cause and effect, and denies the separable nature of reality.

As to Entanglement...those who relate its action solely to the speed of light, will renew claims that "all is written". They will believe reality has no separable nature.

It will seem to be proven, to some...that man is machine, without freewill.

...

Those who feel that Consciousness is the Ground of Being... this saying cause and effect is an allowance of immediate dissemination...will see that things in nature are separable.

C.

What of the seers sight?

A seer enters a trance-like emptiness. There is a clearing away of possible interference from human discrimination... this image inclusiveof the oracle at Delphi, Edgar Cayce, and others. This could include you and I using the I Ching, or similar means of divination. (with the I Ching you don't see happenings, you encounter the tenor of an event or place)

A seer sees between thoughts...and while nearly arrived at the other end of the pathway.

How a seer sees depends on "what is time and distance?"... not only of this place but of a larger reality. The seer depends on the "light in the darkness" of existence beyond what we know of.

To relate events solely to the speed of light is to make it difficult to understand how a seer sees, and a man has freewill…at the same time.

I should say "added" difficulty, because man has, obviously, wrestled with fate for centuries.

Also, to see reality as a carpet, with events relative to Consciousness, is to introduce wrestling with that beast.

…

That there is fate is accepted by most. Acts of "seeing ahead" seem to occur. And to this point in history, it has been easy to depict and to accept the mechanism of the event, without needing to understand it. But, now, Entanglement's suggestion that all things are connected…are of a "oneness"… could bring incorrect assessment, and add certainty to ideas that try to bury freewill.

…

For the Greeks…it seems the Oracle's words were attributed to Zeus. And through centuries the works of "foretelling", although wedded to fate, have usually been seen as resting under Devine control.

Who can blame man for wanting that bit of comfort?

…

Many physicists are saying "It's time we paid more attention to the nature of Consciousness.

…

There are those who are saying man made everything by thinking about it…and is now doing so.

To me…if this universe is part of a larger reality, the moon is there according to laws allowed by the Ground of Being. Our moon…with parts of pure consciousness, is where it is according to…its…best existence, not any man's.

• • •

The moon…and…yesterday, today, and tomorrow… "becomes" in this pace, according to Consciousness.

According to the condition of Consciousness, tomor-row already is present, in the timelessness of an Entirety's becoming the "same as" and "different from".

This is how seers see, perhaps…because they are experiencing a nearly full sharing of the "I" of it all. They see the event ahead. But…also…according to the speed of light in this universe…tomorrow has not yet occurred. Reality is dealt with as one thing following another. This assures us of having choice at the bridge. If I reach a sharing of "I", to a deep point of openness, I will, perhaps, experience both realities. I will see tomorrow, if only in the tenor of it, and will live today, if only in the imperfection of it.

There is fate.

There is freewill.

Both are true.

PART 14

Trekking the rocky road

A.
Imperfect according to necessity.

At the end of reality's road, if all rests whole, complete… as to possibility, we travel with the com-pleteness in each part. This as…the man at his end is in the child. And yet, what I say of imperfection stands.

The reason we move onward is because we seek what's missing. We love. We need.

…

Some believe the world is "becoming" due to man's observation. I think the dreamer of it is Consciousness' "I". As sharer's of that "I", we have the responsibilitythat comes with being able to choose our path.

…

We do our little part by trying to be clear.
While clear…if we love, we honor love…the "I" of it all.

Love includes justice...and swimming hard to keep from drowning.

B.

The stuff we're made of.

The Ground of Being is the stuff of this universe, and of all existence. We need to be careful of the unques-tioned certainties that can blind us to additional knowledge.

...

It will be awhile before we know how a seer sees.

Now...most of us think we can sometimes see tomorrow. We believe in fate. Also...most of us think we're not machines; we possess freewill, when choices are made at the bridge.

C.

The simplicity of "best path".

If one is "loving only"...even in seeming chaos, perhaps that of a raging sea...one is at least hearing the true music of his fate. His mind has been swept clear of all but the love of "what is". So...he wants to keep the "what is" part, as he tries to do what is there to do. He swims hard.

D.

Fate and freewill.

There is fate.

There is freewill.

When our path is cleared, and we are almost fully with the basis of existence...being of the stuff...we are within love necessitated by imperfection. So…in our usual state of mind, perhaps we should see that even if love is hidden, love prevails. Why not get on that street-car? Easier said than done, most would agree.

Reality us "becoming", but sometimes it feels as though we don't have a dime for the streetcar.

• • •

I am allowed, within cause and effect…of the speed of light or the condition of consciousness…to be "this" man and no other.

• • •

As to "becoming"…permanently clear, I am Buddha, Christ, Mohammad.

Occasionally clear, and briefly…I am every man trekking the rocky road.

AWARENESS…

Afterword

Perhaps…

The "Something" which is "this existence" entered, and continued by immediately disseminating itself as laws. The embodiment is "Here" everywhere.

"Consciousness and love…are they the basis of it all"? asked Millie.

"Millie…entangled bit of matter, worlds apart, instantly co-respond if you touch one. For those who believe nothing can exceed the speed of light, there is no message sent between the bits. This erases the separable nature of reality… with cause and effect. It says you and I are without freewill."

"My man, do you have a bullet faster than the speed of light…to defend humanity, and our love?"

"Millie, I believe cause and effect stands. There is message sent. Time and distance are relative, also, and primarily, to the immediacy of Consciousness."

"Drop your pants!" she said. "It's already Tuesday!"

AS TIME GOES BY

Alfred John Dalrymple

JUST ABOVE NAMCHE

(before an unpainted house)

Alfred sits at a weathered wooden table, in a patio, this overlooking and within a high-mountain valley which stretches away and downward. A river, hidden below, shows briefly a few miles distant as the valley disappears southward.

A wood railing helps enclose the task at hand, as Alfred reaches to papers on the table.

From behind him there enters a clear, youthful female voice. He turns his head.

Female (near building)-

I see you have company. I'm here for the porter job.

Shall I leave…return later?

An older female nearer to the table has entered the yard from the adjoining one to the west. She is on the Namche trail and has come up from that town.

Older female-

The say you're mister Dalrymple. I'm Barb Stanton, here because you want a book brought down to the internet café…and put into Kindle.

I'm with the Texas expedition…stuck in Namche a few days.

Will you ask the lady, inside, how many rupees for a cup of tea? (she comes to the table and sits)

Alfred-
She doesn't sell tea to passersby. This is her home.

Barb Stanton-
I need tea.

Alfred (to the girl near the house)-
Are you Sherpa...and at least eighteen? Wait on the bench, dear. I'll be through in a minute.
Again facing the woman at the table, he offers his hand, but she doesn't take it.

Barb (looks toward the papers)-
I assume you have the book on disc. How many pages is it?

Alfred-
It's an essay. Seventy pages. Yes, it is on a disc.

Barb-
I'll take it...and put it on for two hundred dollars.

Alfred-
That's more than I want to pay...Barbara.

Barb-
That's Barb! At the university I don't do these tasks.
Others do them for me. I can use the extra money.

Alfred-

So can I. Thanks for coming up. See that small building... just as the trail cut downward? She'll sell you a cup of tea for about ten rupees.

Without a word or nod, Barb begins to depart in the direction of the building he referred to, but after a few steps she turns and speaks.

Barb-

I think you'll find that the Nepali won't know how to do it. They'll charge you five bucks, and you'll be awarded a boobie prize in Stockholm.

Alfred-

For me...that would be a step upward.

After Barb departs he sees that the girl near the house is sitting, looking downward, and is holding her head.

Alfred-

Do I know you, dear? What's your name?

Girl (she moves forward)-

Calling me "dear" would be more appropriate if

you were older.

(she comes to him)

My Sherpa name is Jangmu. You can call me that, or Nell.

My western name is Nell Barton.

Alfred (offers his hand, and she takes it)-

Is it all right if I feel it?

Nell-

Feel what?

Alfred-

The "dear" part.

Nell-

Why do you want a female porter rather than male... Alfred?

Alf-

Call me Alf, please! I would choose a girl porter only
If I like being with her. I never did have one. And why
not? I'm a man. I...

Nell-

I'm not a prostitute.

Alf-

Nell, you sound sixteen...in the tone of your voice. I
see you're not...but of course that's for you to say.

Nell-

That's right!

Alf-

You have Sherpa features...in the softness, and
gentle curves. You're the prettiest one I ever saw.
Nell...I study you because I've seen those eyes
before...in New Hampshire. No?
Maybe I saw you in Namche a few years ago.

Nell-

My mother is sherpa...my father comes from New Hampshire.

Alf-

Nah! It can't be what I'm thinking! Are you...? No, of course not! But...one day in Concord, near the Capitol building...I saw a pretty girl sitting, looking down, seeming to be in despair. I asked "Can I help you?"

Nell-

That was you?

Alf-

You said you were all right. So I went away. What wrong?

Nell-

After a few years at school...I suddenly wanted to be here, and was having thoughts of a happy childhood.
(she shuffles her feet)
Can I sit?

Alf (pushes aside the papers)-

Would you like tea?

Nell (sits)-

Yes, thanks. What are the papers? In Namche, when they said a man wanted a female porter, they smiled, and winked. Then I heard your name. I read a couple of your books. Alfred John Dalrymple. Isn't using your full name a bit pompous?

Alf-

As a kid I attended at least a dozen schools. I used my first name and my last...with the two d's together. And I was often asked to say it two or three times.

Nell-

Why not just Al Dalrymple? But…now I'm being insensitive.

Alf-

When I write a book I use my full name. The world can take it or leave it!Screw it, Nell!

Nell-

My name is Nyellie Trululu Barton Jangmu Sherpa.

Alf-

It sounds like an ancient Scottish name, that probably never was…but someone dreamed it. The Nyellie Trululu part.

Nell-

Yes…I've thought that, about the Scottish dream. My nutty grandpa gave me the name…and died the same day. Then he went back to the highlands…probably.

Alf-

I have Scottish roots
(reaches for his papers)
I'm wondering if you read my last book, about Consciousness being the basis of existence.

Nell-

Yes, I did.

Alf-

Did you like it?

Nell-

Yes. And I love your novels…particularly when they happen here. They're very romantic.

Alf-

I need to finish my essay. I'm glad the other lady was overpriced, because I need to review this. Can I read it to you?
How much do you charge as porter? I'll pay you more to listen to this…and then be a porter.

Nell-

Six U.S. dollars a day, for three days. If the trek is longer…four dollars a day. More than ten days is three bucks.
(she looks toward the railing)
In Namche, my mother runs the uppermost hotel on the Everest trail. The one with the long courtyard for tents. There's an outhouse at the end. I'll show you.
(she gets up)

Alf-

I know the place, Nell. A dozen years ago I was there, but it was the first week in March, and cold. I was the only trekker, except for Bhim.

Nell-

I know Bhim. He was your guide?

Alf-

Yes. We ate in the kitchen with the ladies of the house.

Nell-

Alfred…you know my mother and probably a couple of my aunts…and my sister, Rita. Not me, though…I was in New Hampshire.

Alf-

I'll read my book to you here. It'll take only today and tomorrow. Ten dollars a day for that job…then we'll trek up to Lobuche. I might go to Changri La, but you don't need to.

Nell-

Shangrila…?

Alf (corrects his pronunciation)-

Changri Pass. Have you been there?

Nell-

When I was ten we were tending sheep above Lobuche, one summer…and we went west of Kala Patar, toward Changri La. But I didn't see much, because it was rainy and misty. And we never went back to that area, anyway.
(looks at her watch)
It's noon, Al.

Alf (touches his papers)-

Can I begin reading this to you? Are you ready?

Nell-

My mother was expecting me to help at lunch. I can be back in two hours.
(gets up)
Is it all right?

Alf-

So…you're about twenty eight?

Nell-

I'll get my trekking stuff together, Alfred.

Alf-

Will you have lunch below? If you don't, I can have something for you…and tea.
After she departs he goes inside…and stays awhile. As she returns, he sees her through the window, and goes out… to the table.
She has a large nylon packsack, which is limp, apparently half full.

Nell (gives him several photos)-

It's just me when I was little…here in Namche. I have some of New Hampshire, I'll show you another time, maybe.

Alf-

I recognize your mother…and maybe I've seen your aunts. This one is sister, Rita, when she was about fourteen.
Is this you?

Nell (looks over his shoulder)-
Yes…I was seven.

Alf (puts two other photos beside it)-
And these…?

Nell-
I was three in this one. Five or six in the other.

Alf-
Could you talk as fast as Rita?

Nell-
No!

Alf-
You didn't need to…you just smiled.
(he studies her)

Nell-
What are you looking for? Can I blink?

Alf-
Your eyes are greenish brown.

Nell-
Look deeper! Maybe you'll find something you lost.

Alf-
I have a large pack, Nell. In a couple days I'll be in shape…then I can carry most of my stuff.

Nell-

You didn't ride to Jiri...and walk to Lukla?

Alf-

I flew to Lukla...spent a night there, and in Phakding. Then up to Namche. I need to walk another day to feel physically comfortable.

Nell-

My sleeping bag will be on the bottom, in case I need it. Usually I'll get a blanket from the people at the places we stay.
Put your cold weather stuff above my bag. Do you have boots? What about a warm jacket and gloves? You should have down pants.
(points to his bag, leaning against the table)
You can carry the things you'll need tomorrow.

Alf (picks up a few sheets of paper)-

Are you ready to listen? Call me Alf...when you're commenting. Can you say it?

Nell-

What?

Alf-

My name. Pretend to be critical, Nell.

Nell-

Al...what you said is ignorant. Did your mother's milk have laughing gas in it?

Alf-

I guess that will do.

Nell (looks at her watch)-

I'm ready as I'll ever be!

Alf-

All ready…that ready? Really?

Nell-

What is time, Al? What is mass? And…do I have freewill?
My education allows me to know I'm ignorant.

Alf-

We're both ignorant, Nell. But…we'll be sincere in our imagination and use of logic.
Nell…I think consciousness is the Ground. All things come from it…including mass and its necessary laws.
This universe is part of its functioning.

Nell-

You feel that the laws existed…before we banged into being.

Alf-

It's why the singularity existed.

Nell-

Time and space didn't start with us…in your opinion.
We didn't come from nothing…but from something.

Alf-

Nell, I think whatever we came from, missingness occured in it. Lack entered, recognized as lack and, so, became need. It followed that whatever had the need wanted to continue to have it.

Nell-

Need? Are you speaking of love?

Alf-

Nell, also we're dealing with imperfection. It's what love's definition contains.
Whatever loved…found a way to continue to do so, as energy effecting.
I believe energy equals mass.

Nell-

You mean in all reality.

Alf-

It has been shown, Nell, that in temperature almost zero…there seems to be unimpeded flow of whatever energy you can imagine.

Nell-

What is time?

Alf-

In this universe…as it serves our experience, Nell, it's one thing following another, as we relate to the speed of light.
In all reality…which includes us…time is an aspect

of Consciousness.
It is the change occurring in the continuation of Consciousness, as it goes forward the "same as" and "different from".

Nell-
From what?

Alf-
From itself, within each unit of the movement. The "same as" and "different from" enable consciousness to be of "twoness"…a necessity of laws.
The resulting unit of continuance occupies space which is "other than" the last space. So…"other than" allows time and distance.

Nell-
I don't understand size in this movement. Is it the smallest parts…as in quanta?

Alf-
I think so, Nell.

Nell-
Imperfection is love. This needs that, onward. Consciousness radiated, or rather, it effected and effects "other than itself", from itself.
How could it manifest what we observe?

Alf-
"Our Something" is imperfect. Consciousness disseminated itself immediately or there would have been no continuance of the imperfection.

The laws needed to be established as being of the Entirety.

Nell-

But…Al…how was consciousness able to do this?

Alf-

I use my logic and imagination, as a blind man putting forward his cane.
I believe mass is energy able to effect.
Think of a boson field in space. A boson is a particle with no discernible mass. It's said to have non-zero mass because it effects other than itself. If particle "A" passes through its field, the mass of "A" can be Increased, as its "spin" changes.
The energy of the boson changes movement, and mass increases. It could be by a slowing effect.

Nell-

Consciousness effected itself forward. My logic can embrace that.
As to bosons effecting, they say we haven't directly seen ninety five percent of this universe.

Alf-

It's acknowledged to be there because it effects.
So…it is energy.
(picks up a paper)
When Consciousness continued…twoness was enabled. And…laws came into being.
We can have moveable and meaningful, manifest reality…only with laws.
In my last book I said reality is imperfect necessarily.

Nell-

That's because love is the "isness" of this mother.

Alf-

If love is..."Our Something" is
Thanks, Nell.
Without Consciousness wanting to exist...and, so, manifesting itself...there would be no purpose in "change", or movement.
Consciousness effected as movement...which is energy. Energy equals mass.

Nell-

Al...tell me another part.

Alf-

In temperature near absolute zero...I'll call it non-zero...there was no resistance, perhaps, to the flow or movement of that energy.

Nell-

I've read about experiments at extreme cold. I could teach your course. Nah! They'd call me Nincompoop.

Alf (with paper in hand)-

In this universe, time is one thing following another, as we relate to the speed of light. For all reality, time and distance are relative...also, and primarily...to the immediacy of Consciousness.

Nell-

Al...I appreciate your effort.

Alf-

Thank you, Nell.

Nell-

But…maybe I'm thinking "Al is a nincompoop!"
(she hits his arm)
I'm kidding!

Alf (points to his papers)-

In this book I'll try to show that "now" has three parts…and they occur, to our direct discernment, as one part.

Nell-

Can I please have tea…and a cookie?
(stamps her feet)
Dammit!...I don't need a cookie!

Alf-

Nell, please go in to aunt Mingma…and get us tea, and a pack of cookies.
(reaches into his pocket)
Here's three hundred rupees.
Each note is a hundred rupees. She takes two with her into the building, and soon returns.
She puts two cups of tea, and a package of cookies onto the table.

Nell-

We can have a coconut macaroon.

Alf (as he opens the package)

Do you mind if we go on, Nell? I need to get half of this done today.
Take a handful of cookies. How many are there?

Nell-

Fifteen.

Alf-

Take eight.

Nell-

Three is my choice.
(takes three)
I'm ready, Alfred.

Alf (again points to the papers)-

I'll try to show that, at the Ground, "now" consists of three parts...due to each unit of reality going onward the "same as" and "different from".
Even as we speak each "now" is the "same as" and "different from" the last one and the next.
If time is consciousness "becoming", twoness was enabled...the continuing units of which are the "same as" and, so, connected...but also "different from" and, so, separate.
Every unit of reality goes forward in this way.

Nell-

How do I have freewill?

Alf-

In this universe, where one thing follows another in

a linear way, as to our usual measurements, and we relate to the speed of light, tomorrow has not yet occurred.
At the bridge...you use your "I" to choose. The next step can be made by you, alone...separate from God, genes, or other men.

Nell-
You mean...before I'm interfered with by genes, or other men. Al...I can be controlled. Most humans are.
(she touches his arm)
There is great fear connected to change from early learning.

Alf-
Sure...Nell, this seems very true of the unquestioned certainties. They're old friends. It's why this group wails at a wall, and that one bows to the east.
They deepen differences between groups. "I am better than you!" stands as identity and comfort.
In the Christian world, continents were conquered through masked unquestioned certainty: "My God is better than yours!"
Let's get beyond this!

Nell-
That? Or...do you mean this moment, which rests on Alfred's certainties?

Alf-
Ah, Nell...we agreed we're ignorant. But we're trying not to be.
Will you listen?

Nell-

Yes, Al…I'm listening!

Alf-

Our parts are made of smaller parts…down to almost "not there", it seems to us. They not only obey laws, they are the laws.
This is true if energy equals mass.

Nell-

We're getting down to the Ground. It's at the end of your tongue…even as we speak. Sorry!

Alf-

Nell…why are you angry?

Nell-

And yet…even at the bridge there seems to be fate. We also go forward connected.

Alf-

At the bridge…if your mind is clear, "best path" will call. That's all I want to say, now, about fate. Later, we'll speak of it. If you approve…I'll roll it off the end of my tongue.

Nell-

Al…to be at the bridge properly, I need to accept my aloneness, and be clearheaded?

Alf-

Your separateness…yes…and likely there'll be music calling. If you're open.

It might be a brief experience, but you can be there fully engaged by what is before you.

Nell-

I went to a concert, and heard Rimsky-Korsakov's "Scheherazade". It seemed that my self was gone into it. I was fully there.

Alf-

Good, Nell. That's what I mean by "loving only". without acknowledged judgment your connection was free of the worst parts of human mentality.
Unfortunately, Nell…most soon return to the usual discriminations, and lose the clearness.
But…"hearing the music calling" is a phrase which holds a universal theme. Something is beckoning to you. I think the call is from "best path".
You can't be so full of yourself that your ears are plugged.

Nell-

Al…I know what you mean. It's Bali Hai calling, as your special place.

Alf-

I think so, Nell.

Nell-

Al, it seems that most things…from the smallest parts of atoms, to rocks, trees, skin and bones… don't have "I".
What of them?

Alf-

All things without "I" maintain themselves as they follow "best path"...according to the laws of continuance. They go on existing because they are the laws, in the sense of being part of the Ground. You can say they have pure consciousness.

Nell-

That's why they're smarter than we are? Sorry! But, the atom does what it's supposed to do. The ant leads its unquestioned life.

Alf-

Material ethereality goes on due to imperfection... which we can think of as need, or love. This needs that. It's why the "same as" and "different from contain the twoness necessary to the spin a non-zero mass boson can impart. Love says "this" won't spin left unless "that" spins right.
Existents without "I" of their own, share the freewill of the "I" of it all...this due to laws.

Nell-

Our own body parts don't have "I". How far down does it go...this pure consciousness?

Alf-

Down to what maintains itself. To consciousness going on as laws.

Nell-

I won't say "sorry!" this time. Why are they smarter than we are?

Alf-

Nell…only creatures with "I" can choose "worst path", as opposed to his own definition of "best path".

Nell-

Ignorance seems forgiveable, Al. Must I be alone?

Alf-

I think…that's only in your choices, Nell. You're separate when making them. But you're not alone, in the sense of loneliness. You can love and be loved.

Nell-

At the bridge I ought to be clearheaded?

Alf-

Your pond shouldn't be so full you can't let in something new.

Nell-

I can see the other side of the bridge more clearly, by avoiding putting thoughts you would call "crap" between me and it.
But some are so lonely…they stop listening.

Alf-

Nell…you seem very caring. Have you ever been Married?

Nell (pulling away slightly)-

A marriage was arranged when I was a child…but my father interfered. When I was in New Hampshire, at school, I was proposed to, and then you saw me holding my head. I came home soon after that.

Alf-

A pretty girl like you…alone?

Nell-

With friends, at elbow distance. Wondering where I belong. I'm Nyellie Trululu Barton Jangmu Sherpa.

Alf-

Doesn't love provide the place, Nell? Love is Bali Hai calling…and you're suddenly where you want to be, doing what you want to do.

Nell-

I wouldn't know!

Alf-

Nor would I.

Nell-

And, so…shall we move on, Al?
Your last book spoke of "entanglement". You said two bits of matter can interfere with one another… and, after that, seem to act as one.
The bits might be very far apart, but co-respond when one is stimulated…you said.
Poke one…they both jump? For some…this erases

the separable nature of reality. I know the matter is important to you.

Alf-

If things are separate we can have cause and effect.
A thing jumps because it has been effected.
Time and distance are relative also, and primarily, to the immediacy of consciousness.
There is message sent between the bits.

Nell-

Al…at the bridge, some think our next step is determined by genes or by the kind of fate which is "written". And you think this would make us robots.

Alf-

A machine's actions are of a program other than one which is self-determined.
We're human…our "I" is separate at the bridge.

Nell-

If we're free to do so, Al? But…I'll put that aside.
As to movement…isn't the distance between the bits also non-zero?

Alf-

That has to do with the carpet's immediacy, Nell. We obey the linear…but, also, the distance is due to "change", as consciousness goes on the "same as" and "different from".

Nell-

You can continue with your reading.

Alf (reads)-

Love is imperfection. This needs that. Twoness is necessary even to the spin of a particle. This won't spin left unless that spins right.

Nell-

Explain "spin", Al.

Alf-

Tiny particles, even when seemingly at rest, have attributes of a turning object.
Add an electric charge, Nell, and much of matter becomes magnetic.
Where was I? Oh, yes…the laws of nature need "twoness"…as time goes by.
(he laughs)
Don't touch my finger!

Nell-

You sly devil!
(she blushes…and clears her throat)
That aside…what is the purpose of our existence?
To make consciousness happy?

Alf-

You and I are that part of Consciousness…sharing "I". We don't have purpose as duty…we have it as "love". And we can say why we're happy.

Nell-

I asked you a question, mister Dalrymple. What is our purpose?

Alf-

To be happy!

Nell-

That's too simple. Also…it's so hard to do. I might stand at the bridge…indecisive…as an aunt walks by on its "best path".

Alf-

The purpose of most things is to continue according to the character necessary to the continuance of that thing. They maintain their character. Thus the ant.

Nell-

I'm not an ant.

Alf (reads)-

You could say…during the first action of consciousness, love was the mother of it. Hereness, according to love, needs thereness. This needs that. I need you.
Of course…I don't.
Any other questions, Nell?

Nell-

How did Consciousness create the first mass? A quick picture, please!
(looks at her watch)
And I don't have forever!

Alf-

How would I know?
(he sighs)
"Our Something" had missingness…it followed that

love loved itself. This awareness moved in order to continue. Energy equals mass.

Nell-

I like that, Al…but…isn't it too easy?

Alf-

I'm ignorant! I look for the bottom line!
"Our Something" knew it was and, wanting to go on, moved to effect the "same as" and "different from"… as to its own seeming oneness. This energy enabled the twoness necessary to laws.
Considering all reality…I think the only meaningful "oneness" is love. It effected to be all imperfection. Consciousness is love.

Nell-

Probably. But…I spit on your certainty!

Alf-

At least you didn't spit in my mother's milk…with the laughing gas in it.
(points to his papers)
Are you ready?

Nell (laughs)-

I'm ready!

Alf-

In all reality each "now"…you can call it an instant… is the "same as" and "different from" the last and the next, or there would be no existence.
Every "now" has three parts, both separate and

connected. This allows freewill and the possibility of seeing at least the tenor of tomorrow.

Nell-

You don't believe in fate?

Alf-

I do…but my logic won't encompass the usual understanding of it.
Nell…freewill must dominate fate, or we wouldn't be human.

Nell-

How can a seer see your buttcheeks being cut off by Achilles, at that next battle?

Alf-

I need to think more about that. When the seer is betweenthoughts…in a trance…time is outside our usual limitations as to where our "seeing" can be.
Let's not discuss fate! I want to talk about "now".
Sure, Nell…we have last and next, but think about "now". Is there "now"?

Nell (she stands)-

I'm here!
(she takes a step)
Now…I'm here. Time, which includes the "now"…passed.
And…here I am!

Alf-

Nell, a game that takes three hours to play is on "now". During the performance, I could've phoned

you and asked if they were doing it "now", and at any instant you could have said "Yes"...but you would have been merely accommodating man's perceptions.
Each instant during your step you were the "same as" and "different from", but man can't discern the three parts. We can't see movement, or the parts of "change", as consciousness goes onward.

Nell-
So what? For all practical purposes we see it. The technical gets lost in the experience. Useable truth is dominant.
(she gets close to him)
See my tongue yet? I'm baring it toward you..."now".

Alf-
Why did it take so long to get it out of your mouth?

Nell (sits)-
Your bottom line is that your opinion is a credit card good anywhere you want to use it. If you want to buy my attention, how much credit do you have?
Are you bottomed out yet?
(she laughs)

Alf-
Ah, Nell...don't be angry!

Nell-
You're trying to understand time. Most others just live it.
(she smiles)

Don't be so alone at the bridge, you say there is no "now" on face book.

Alf-

Nell, I'm amazed at what pops into your head. It's always pointed.

Nell-

What is?

Alf-

Huxley and Orwell agonized about minds being controlled...both depicting man as manipulated and machine-like. And Orwell showed every household as having a camera in it, relaying all action to a central office. He said "Big Brother is watching you!"
Isn't this, to some extent, facebook and its relatives, telephone and email?

Nell-

What are the consequences?

Alf-

Perhaps one of them is this:...with the loss of privacy's freedom, comes the loss of friendship. Nell...my friend knows when I say "Rob a bank!" I mean "Play golf!". If a stranger intercepts the message, I can arrested. So...knowing "Big Brother" is watching, I'm not likely to speak freely to my friend.

Nell-

I shudder to think of the death of privacy. It's my

choice to be a nincompoop when I want to.
(she shudders)
We are entering the age of robotism, Al. I understand your concern. They'll be putting chips in our heads, to suit their definition of "happy".

Alf-
I wonder if it can be stopped.

Nell (points to a group of five children, ages about ten, eleven, twelve, who have appeared from below and are walking through the patio. She glances at her watch, then speaks to them)-
Don't be late for school! Don't stop to play in the forest!
(she raises her arm, and waves a finger, in a scolding manner)
Have a nice day!

All five smile and bow, as they continue on. They each have their hands clasped high, before them.
(they shout)
Namaste!
Now they go from sight…around the edge of the building.

Alf-
Did you walk through this patio to school, Nell?

Nell-
Yes, Al…up to here, then through the forest to Khumjung…to the school Hillary built. Also, he Helped build a clinic there.

Alf-

How many years did you do it?

Nell-

Ages ten and eleven…before my dad took me away.
With my mother's help, he did. She walked with us to Jiri…and rode the bus eleven hours to Kathmandu. We said goodbye at the airport.
I returned at age twenty three…and stayed a year.
Al…when I was at school in New Hampshire, one of my professors posed a question…and it's the only one I remember him asking. All he ever did, it seems, was open a book and read from it. But…suddenly, one day he raised his head.

Alf-

And…? What question, Nell?

Nell-

What would you do…if you could do one thing to Change man's condition? So…?

Alf-

So…what?

Nell-

What would you do?

Alf-

I don't know!
(scratches his head)
The professor seemed worthless, but maybe he was

bored. Let's begin with what you remember...then five children walk by on their way to school.

Nell-

The professor didn't give us his answer to the question. Is it because he lacked the courage?

Alf-

Maybe he was waiting to be asked.

Nell-

Shouldn't he have given it? For me to admire him, what should he have done other than raise his head one day? Looking ahead...who do you suggest I do business with?

Alf-

You have your own way of deciding the value of things. But...look for someone who is smart and kind. Nell, the man who understands he's made of the stuff of the stars...at least tries to see that each center has the same innocence. If he knows it with his head, he ought to, also, feel it with his heart....or try to.

Nell-

If the professor felt it, he could have given an anwer to my innocence. Why not assume some of us have, or will have, our own mind?
Al...he should have asked himself "How can the world be served?"

Alf-

Do business with the one who asks that.

Nell-

I am! But there's something bothering me.

For awhile both are silent. They sip tea, and study the mountains.

Alf (sighing)-

Nell, even as I apply logic to fate...and give my
answers...I feel that I don't know much.
If we have freewill at the bridge...make choices
there that are separate from genes, other men, and
God...fate is subservient to that occurrence.
In this universe, tomorrowhas not yet occurred...
thus, due to the separable nature of reality, our
choices determine what comes next.
But, because we are allowed to go forward "different
from" while the "same as" continues, there is
connection between now and later, as on a carpet.
Those able to enter a trance-like condition can,
perhaps, see ahead. The oracle at Delphi, as depicted
by Greek thinkers. Nostradamus and Edgar Cayce.
Because of the nature of time in all reality...perhaps
they saw results of what freewill brought about.
Ah, Nell...I don't want to think about it now.

Nell-

I feel that they can see ahead, Al.

Alf-

It seems that events ahead are determined by acts

of freewill along the way. Freewill determines what will be seen. And…yes, I do believe seers can see ahead. But I need to think more about this…later.

Nell-

In mist, if I see Achilles cut off your buttcheeks at tomorrow's battle…you must have chosen to be there.
You wouldn't hide in the outhouse?
(she smiles)
I say that…because I know you would go and fight.

Alf-

Maybe I'd lop off "his" buttcheeks. But we should apologize to Achilles since we never did meet in battle, even though we're only making a point. I think you're right, if you see me there, it means my choices got me to it.
Goodbye, my buttcheeks! Anyway…I'm not a coward.

Nell-

The ladies at Delphi know you're not. They see Alf's cheeks being endangered.
(she reaches to his hand, and briefly holds it)
I know you would go to the battle, if you thought it was your duty to.
Let's move on!

Alf-

Fate remains a mystery to me, except to this point: in our universe freewill is dominant, or we couldn't make our own choices.
I am not a robot.

Consciousness precedes DNA.

The woman of the house has entered. She approaches the table.

Aunt Mingma-

I have Chapati and jam.
(she puts it on the table)
I'll bring coffee for you...and apple pie.

Nell (to Alf)-

The apples are from Jorsale. The coffee is from Mustang Province...and it's very good, but strong.

Alf (to aunt Mingma)-

Yes, please! How thoughtful of you!
(he reaches to his pocket)

Aunt Mingma-

You don't need to pay.
(she departs)

Nell-

She likes you.

Alf (as he puts jam on the flatbread)-

Nell...this early in September, will there be a trail through the snow at Changri Pass?

Nell-

Not likely. Probably no group has been there yet.
When we get to Pheriche I'll ask my aunt Kamala.
I have no relative above that altitude.

Aunt Mingma reenters carrying coffee and pie.

Aunt Mingma-
I've been thinking about you.
(she now is busy with the coffee and pie…placing it properly)

Nell-
What did you see, auntie…when you thought about us?

Aunt Mingma (sighs deeply, smiles, then grasps the right hand of each, and joins them)-
The shorter route is in darkness. I saw blood, and heard echoing sighs, crescending…is that a word? Echoing…on to a big bang.

Alf-
As we climb to Changri La?

Aunt Mingma-
There were voices intruding…and the way was slippery.

Nell-
We were sliding down, aunt Mingma?

Aunt Mingma-
Looked like you were sliding up…to me. It was dark and misty.

Nell-
We shouldn't go there?

Aunt Mingma-

Are you crazy? You would stop living because of a bit of blood...and darkness? Keep your wits...now and then!

(she departs)

Nell (shouts)-

Is this one of the "nows"?

(she turns to Alf)

We can get on with your book.

So...what would you choose...if you could do one thing to help man?

Would you take children from their parents? Not just some...but all.

Alf-

No! Even in the worst society, family is where love can be deepest, for children.

Nell-

I agree! Sorry I asked!

Alf-

Huxley and Orwell feared that babies would be born into motherless laboratories.

What would I do, Nell? I would build more clinics and schools. In the schools I would have it said that love is what we're made of. Love is why we're here...why we exist.

Nell-

More words?

Alf-

And when they understand we're made of the stuff of the stars, they study that stuff.
They'll see that things maintain and go on according to "best path". The existence of all things is keyed by love.

Nell-

Al…what about the adults in the "now"? Most of those who control the world would nod their head to your suggestions, smile, then go about the usual business.

Alf-

I agree, Nell.

Nell-

Man…inclusive of the adults in the "now"…must change monumentally, to love properly. How does that come about?

Alf (laughs)-

Kill some adults?
Sorry!
Ah, Nell…with both children and adults included…we study the results of unbiased scientific inquiry, as we get it from the laboratory or from nature. And…we avoid the use of unquestioned certainties.
We'll think more about this later. Can I move on?
I'm not avoiding your question.

Nell-

Considering man's condition…how to make it better

is a depressing question. Yes, Al, let's move on.

Alf-

Nell, the sense of mystery…is one of our greatest possessions.
As a philosopher…I'm only trying to understand. I think freewill and fate are both true, with cause and effect a continuing necessity within the movement of it. But…what do I make of a lurking sense of "destiny"?
What of other dimensions? We speak of feeling something in a "just there" kind of way. And when I left my body I seemed to be just there "here"… whatever that means. I…

Nell-

Now tell me what love is. But…of course…I already know.

Alf-

The nature of love is possible through imperfection. We lack…we need. Now I'm relating our "I" to the "I" of the entirety…or God. The only imprisonment is in the areas claiming to have perfection….the "I am better than you!" places.
True love is trying to see things as they are…and being just and kind when you have to do with them.
What would I teach? That in the body and body-mind, and at the center of "I"…our soul…man doesn't only possess the innate capability to love, he is love.
Ah, crap! What a useless mouthful! Even as I speak, I spit! What I said isn't likely to be seen, and maybe it shouldn't.

We advance through adversity…why get rid of it?
(scratches his head)
Of course, it's understood we ought to get rid of more and more hurtful thinking and action.
What would I do?...maybe nothing! Does the question suppose the answerer doesn't feel helpless in the face of it?

Nell-

Now tell me what you think love is…as a man, not as a philosopher. There's something more than loving what you think is right, or good.

Alf (sighs)-

Maybe it's the sun coming through your hair.
I need to read this!

Nell-

Read it! What is my hair to you?

Alf (reads)-

"Now"...is the connection between the present instant, the last instant, and the next.
"Now" consists of all three units.

Nell-

How is your "now" connected to entangled bits bits separated by vast distance?

Alf-

Message is sent between them. Time, distance, and size are defined, also, according to an area beyond what we know. Light's speed is exceeded because

it's also true that time is not linear, distance is not vast, and size is not here…then…there, but is here… and…there.
Let's skip that for now, Nell.
If Consciousness is the Ground, man ought to study time and distance at the root place he continues to exist…from one instant to the next.
"Now" is the past, present, and future…the last instant, this instant, and the next. There is no "now" except in non-zero distance and time.
At the Ground…the "same as" and "different from" occur as "change".

Nell-
Al, we said…for practical purposes there is "now".

Alf-
Sure! But…we need to investigate the truth. Shall we go on?

Nell-
Yes, go on! I'm listening!

Alf-
Every instant occurs because of need. We are born of imperfection…as love.
Love…whose energy allowed mass…enabled laws.
Nell, at the beginning of what is, or what moves… there would be nearly non-zero resistance to the flow of energy. I…I'm out on a limb.

Nell (puts her hand on his)-
I read about study of temperatures near absolute

zero. I know you imagine on the basis of other's sincere inquiry.
Aside from that…I'm wondering…if you go out on a limb, will you dive into a bush?
(she laughs)
I didn't mean that, of course!

Alf-
I don't go out on limbs…in that way. But I'll dive into a bush…after getting the lay of the land.

Nell (laughs)-
What do you think of the land here? Will you have the lay of it, in your opinion?

Alf-
Suddenly I have a warm feeling inside, Nell. And I see myself getting the lay of it.

Nell-
But, Al, maybe you won't get it!
(turns her head)
My father is coming!

Alf (jumps up)-
Where?
(when he sees that he and Nell are the only ones in sight, he sits)

Nell-
He said he'll be in Namche in late September. A few weeks from now. Maybe.
But it could be tomorrow…or in October.

Alf-

What does he do for work?

Nell-

He's a professor in the chemistry department of the University of New Hampshire. But lately he doesn't have classes…he does research, and makes stuff. He's been making robots.

Alf-

Why didn't you tell me about this, earlier?

Nell-

I thought it might disrupt your reading. But I did Intend to tell you. Just didn't get to it. Sorry!

Alf-

Does he seem to care…deeply…about how the robot And the human differ?

Nell-

He does speak of it.

Alf-

Nell, what is a robot between thoughts?

Nell-

Between whose thoughts?

Alf-

As to ours, Nell…between them, you and I are loving, and without hurtful discrimination.

Nell-

You hope!

Alf-

Between thoughts…you and I go on as the purest
part of our souls.

It seems that Nell is about to answer, but she changes her mind. She has a sip of tea, and then another.

Nell-

What is my hair to you?

Alf-

If I loved you…which I don't, of course…your hair
would be more than just another item to love. It could
be thought of as special. I suppose. I…

Nell-

It would be "my" hair.

Alf-

We ought to move past this!

Nell-

Go ahead!

There is some laughter which comes from a window of the house.

Alf (lowers his voice)-

So…we can love for no apparent reason. I can love

you without knowing why. Actually…I can love you and not know, consciously, that I do. And, later, way up the road, I might suddenly know I loved you from the beginning. I…
Do you believe in reincarnation?

Nell-
Yes!...I was Nyellie Trululu…and you were the ass who came through town. Have you seen the movie "Picnic"? In man's history, how many times did a man come to town, then leave?
How few times did the girl follow him?

Alf-
How often was there no ass in the story?

Nell-
The man wasn't only trying to serve himself?

Alf-
Sure!...more than once the father plugged arrows in the ass, and the ass deserved it.
What does this have to do with being a robot?

Nell-
You asked me about reincarnation.

Alf-
Later, Nell, we can discuss reincarnation. As it is now defined, it feeds caste and privilege, as the percentages of those in the mud stay the same. But…who am I to say?

Nell-

An ass, with an arrow sticking out...maybe.
But...who am I to say?
(she sighs)
Where do you live? In New Hampshire?

Alf-

In Alaska's Aleutian Islands. At Unalaska...which is often called Dutch Harbor. But I was born in New Hampshire...at Portsmouth.

Nell-

I was in Portsmouth. One day I walked near the Piscatequa River...at a place they called Strawberry Banke. I imagined Pilgrims being there.

Alf-

Probably they camped where we played baseball against a part of town we called "Puddle Dock". But they didn't like it when we used that name.
Anyway, I remember a narrow, deep inlet into a pond where boats could escape the river's strong current.
This is conversation for another time.

Nell-

Back to robots, Al?
What if you had a boy robot...programmed to say "Dad". And, one day, when you and the robot are walking beside the river...a full-bodied one, with a waterfall ahead...the boy, I mean the robot, falls into the river. Then quickly he is going outward and away, toward the falls. He looks at you and shouts "Help me, Dad!"

Alf-

Thinking I might die if I went after him...perhaps I'd do nothing. He is a programmed machine.

Nell-

Then he calls again. He yells "Daddy, I love you!"

Alf (shivers)-

Why are you doing this to me, Nell?

Nell-

My father made a robot about two feet tall, with the aspect of a boy, and having the voice of one.
My name was put into it...and the boy, the robot, would say "Good morning, Nell. Will you please take me for a walk?"
The reason I'm doing this to you? Al, you know the difference between a human, with separate choice at the bridge...and a robot, programmed for its action there.
The machine doesn't move there according to self-emanated desire. But...can you "feel" the difference? When the boy says "I love you, Dad!" do you feel that the words come from a machine?
Is this why man will go on making robots...because he can't feel the difference?

Alf-

There are other reasons, but making robots will lead to the effect you speak of.

Nell-

Dad said he thinks someday he could make one you

might marry. Then he laughed and said "If man can make a machine act human-like, why bother? Just take what's already made…a human…and make him or her robotic.Why not put chips in heads to control anger, desire, joy, happiness, sadness…the emotions we have.
The he shuddered…as you did a minute ago.

Alf-
Perhaps he sees it coming, Nell, and others will share his feelings. Some scientists will fight to make robots from "scratch", rather than directly alter man's mind.
In either case…man's mind will be altered.
Orwell and Huxley…in their graves…will shudder.

Nell-
When we're visited, the ships will have robots?

Alf-
And who made them, Nell? Throughout the universe, creatures with "I" will have a skull-like house…a special place for the "stuff". And, those having our amount of it…will think on the same level we do. Poor devils!
Because they'll face great linear time and distance, robots will be sent. And when the robots go home, only robots will greet them.
I shudder, Nell. That's why I write.

Nell-
You need to! And if I listen…perhaps others will.

Alf-

But I might be wrong about Consciousness as the Ground of Being.
For us, in this universe, message shows as here then there. But as it accords with all our reality it is, also, here…and…there. The "same as" and "different from" obey non-zero time and distance.
Nell…message sent here obeys both linear and non-zero time. But…

Nell-

But…man…the poor devil…won't care. Most don't ever tune in to those questions.

Alf-

Of those who do…some will want to continue as human and not robot.
(pushes aside his papers)
Tomorrow we'll climb to Tyangboche.

Nell-

How early?

Alf-

Be ready by seven. We can walk a few hours before the sun gets hot.
(he gets up)
To be fair I'll come down to your place…and we can start from there.

Nell (gets up)-

You don't care for my company?

Alf-

I only meant business is done. You can stay. Why not sleep here? We can sip tea, and watch the sun be golden on the mountains.

Nell-

I told my mom I'd help her with dinner. I'll see you early.

Carrying her pack, she walks away. At the edge of the

122

the hill she turns and smiles, and waves. Then she puts her hands together, and bows toward him.

He makes the same gesture toward her.

TYANGBOCHE

After climbing, steeply, on a forested trail, they arrive at the edge of Tyangboche...and face a clearing with a dozen or so buildings around it. Those on the left, at hand, are among trees...this followed by a dominating monastery. To the right is a longish, one-story building.

Alf (points to the right)-

Look, Nell...what is this new building?

Nell-

A hotel. It's the best one for us.

(she walks ahead)

We go to the back of it...toward the hill...then climb stairs. It has a courtyard...with bathroom and shower house. Flat rocks are there to walk on, in case it's muddy.

They go around the north end of the building, to the courtyard, then climb stairs which lead into a large room, obviously for dining and check-in. As they now face south, the kitchen is to their right.

Nell (shouts)-

Aunt Chiri...it's Jangmu.

A tall, slender woman comes from the kitchen and gives Nell a strong hug. As they converse, apart, the woman glances his way several times.
She comes to him.

Aunt Chiri (smiles)-
You can stay here. I'll show you the room now.
(she turns and walks away)

In the room, which is about eight by ten, and of unpainted, locally sawn evergreen, there are two wood-only bed bases, these having mattresses a couple inches thick. He drops his pack onto one, and Nell puts hers on the other.

Nell (removes a few things from her pack)-
I'll sleep in the kitchen, of course.
(looks at her watch)
It's nearly dinner time. Would you like to speak
about your book before we eat? You can come…in ten minutes.
(she clasps her hands near her forehead)
Namaste!
(she and Aunt Chiri depart)

In the dining room again, he and Nell sit at the south end table, beside a window. They can see part of the courtyard. Other trekkers have arrived, some seated near them. Aunt Chiri is distributing menus.

She comes to them.

Alf-

I'll have the Sherpa stew, please. Also…the two-egg omelet and toast. Thanks.

Nell-

Yes, that's what I want, Auntie. I'll get tea for us. (she departs with Chiri)

From the nearest table, about six feet away, a young male trekker speaks to Alf.

Trekker-

Sir…did you come from above? Will we get past the cloudiness…soon?

Young female trekker-

We had the sun only one hour today.

Alf-

The summer clouds linger…but after a few more days it will clear. Don't be discouraged.

As Nell returns with glasses of milk tea, Alf is looking through the window.

Alf (whispers)-

This scene is of evergreens on a hillside you could throw a rock to. But the mist will allow them to imagine things.

Nell-

Tomorrow they'll see high mountains. And it's
Friendly, here...with a shower in the courtyard.
(she sips her tea)
Al...I'm listening.
(points to his papers)

Alf-

I'm tired of it, Nell.

Nell (reaches for a paper, and reads aloud)

Man's observations have brought this thought: There seems to be a "presence" beyond our knowledge...felt to be at hand, or "just there", above.
What can my imagination do?

Alf-

Some physicists are saying it's time to study consciousness. But to apply deep imagination to that subject is to put reputation on the line.
Today...man's conclusions in areas of what could be important knowledge seems to have bogged down in the mud of silence.
(he sighs)
Maybe there needs to be a bit more opinion...blown into paper bags and exploded in faces.

Nell-

Al...some of the study takes place in labs. A bit ago you were concerned about robotism.

Alf-

Sure, Nell…it's another area of consciousness in need of imagination and the sharing of thought. But…

Nell-

Man is afraid to rush ahead the wrong way?
Al…I agree that some should speak out. But…both church and science have sometimes flooded the world with those unquestioned certainties you spoke of…to control the herd.

Alf-

Good, Nell. Let's get back to the book.
What is the consciousness not only "just there", but everywhere? What is the entire basis of existence?
(picks up a paper, and reads)
Almost free flow of energy near absolute zero is an area inviting great imagination…if one is not afraid to say time and space didn't begin with our "big bang".
Increased mass through effect of a seemingly massless boson suggests that energy equals mass, but who want to deny the stranglehold of always relating to the speed of light?
Wouldn't Einstein agree? Man needs imaginative knowledge, even if it's the debatable kind.

Nell-

Sincere insight sometimes raises eyebrows. But we don't hope, if we don't know…at least to the point of "maybe". Did I say that? It slipped off my tongue.

Alf-

Nell...that was a good one.
Man has got to feel he's busy trying to advance in the knowledge of large truths.
Nell...if you observe mass seeming to come from movement, what do you make of it?
If at non-zero temperature you see almost free flow of energy...do you apply this action to why and how we exist?
If not...what are the fears that nail you down?

Nell-

Al...tell me what you think you know, sincerely.

Alf-

If I do my best, Nell, I'll be a mirror to the truth...in the sense of being open to it. I can only take credit for effort.
Humility is the greatest virtue.
What do I think?
Sentience, as Consciousness, resides in all things.
They maintain. They go on. And if the laws allow it... so do the laws go on.
Those existents without "I" go on in a "best path" manner. Every atom, to the smallest part...also, the spin in a boson field...is of continuance. All these share the free will of Consciousness itself, or the "I" of it all.
But they don't have "I".
We have it!
(he sips tea)

Nell-

We have a soul. But speaking of atoms and cells and skin and bones…do the parts of Al and Nell have "I"?

Alf-

No! They maintain according to the laws. They don't do battle with ethics and mystery.
But our "I" is often confused, particularly when trying to see what seems to be "just there".

Nell-

It stays out of reach, doesn't it…except to feeling.

Alf-

But, Nell…for most things in life, it it's a "will-o-the-wisp" we soon know it.
We feel love…and it's "just there", beyond understanding…but it floods the soul…and the feeling of it proves its essence.

Nell-

As we advance through adversity…love can make us both happy and sad.

Alf-

Love allows existence of all fortune…even the kind we hate.
But in the end…someone who loves you will see not just any butt.

Nell-

Moving on…you're saying if there's a Ground of Being, it must be consciousness effecting. To you… this is the energy of existence.

Alf-

I think Consciousness is the only oneness that can make "other" than itself.
It is the only thing that can be "Something which loves its imperfection". It has lack, and need.

Nell-

It "is" love…you said.

Alf-

Consciousness is the only oneness able to make twoness. Energy equals mass.

Nell (picks up a paper)-

In non-zero temperature Consciousness effected "twoness", which has movement, a necessity of laws. All things maintain as on a carpet, bearing micro-cosmic roots.

Alf-

Continuance has action at the smallest parts.

Nell-

I agree that we need to know the roots. Our universe is billions of light years across, but we need to know how it can be of "any" size.

Alf-

Yes...and some despair at their littleness in relation to that universe.

Nell-

And yet...materiality comes from what we think of as ethereal...you say.
Keep that in your book, Al.

Alf (laughs)-

Fortune plays queen to a man's courage.
But...if I'm wrong about Consciousness as the Ground...am I a brave thinker, or am I a nitwit?

Nell-

In whose eyes?
To mine...I hope you're both! I need you to be correctable.

Alf-

Isn't "nitwit" a bit much, Nell? How about just a little stupid upon occasion?

Nell-

I'll settle for that!

Aunt Chiri brings food, and more tea. They begin to eat. After awhile, Nell reaches for another paper, and reads, aloud.

Nell-

If energy equals mass, movement is purposeful. This is acknowledged at the recognition that all things

maintain.
This is good Sherpa stew.
All things maintain, but only those with "I" know their soul is different from the body.

Alf-

Eat...Nell! That last thought of mine is pompous. Other animals don't need to know it.

Nell (not reading)-

If love is the basis of it all, and we're the only ones here that might consciously know it...does this mean that, after enabling the laws, the beginner has no need for a personal relationship with us?

Alf-

Our "I" is part of the "I" of it all as consciousness. When we love, the Ground has to do with us in a personal way. It seems to me.

Nell-

Does the "I" of entire existence, or God, care about me always? I mean does he care about my soul...or the "I" of me? Does anything "just there" or "just anywhere"...give a crap about Nyellie Trululu Barton Jangmu Sherpa?

Alf-

I don't know, of course. In an indirect way...you're loved by the Ground, or God, when someone with "I" loves you. If you believe that, seeking the other answers will be bearable. Perhaps.

Nell-

Am I Nell Barton…when between thoughts?

Alf-

You are loving, there, as Nell Barton. Your "I" is yours only…as "I give my "I" to thee". Of course, I referred the giving…to you, not me. As for me giving my "I" to thee, it would be in pretense, as we sit here having tea. I…

Nell-

Even in your pretense…that's a point of suffering I agree to, and wonder why.

Alf-

Suffering?

Nell-

I think most Buddhists would say that the self is illusory. Each moment I am born, decay, and die. Every instant illusion of "me" renews itself. Nothing is carried over from one moment to the next.

Alf-

Nell…you've listened to me speak of how each instant is "different from" but also the "same as". I think there would be no existence without both. And you agreed, it seems.

Nell-

Yes! And I feel a loneliness. Is it because I need someone to love and be loved by…as though forever? Why would there be personal attendance to "thee",

meaning me, by Entirety "I" or by yours, Al…and have it end at the death of the individual? Of course, I used your name in a pretense kind of way.
We sit here having tea. I…
After meeting in New Hampshire and here, why do we interact as though we're old friends? Is it coincidence, as due to having similar education, and large amount of like character? Or…
But, of course, talk about our relationship is only intellectual.

Alf-

Sure! And…Nell…let's not speak of it.
I didn't say I disbelieve reincarnation. I just think if our souls go on, it might be in a way different from that of man's present understanding.

Nell-

Would you say the "I" of each individual is likely to go on?

Alf-

When sameness becomes apparent, suffering passes because the difference becomes pure. This is a tree, and this is a man…as manifested by consciousness… but, also, due to the movement, or continuance of consciousness…it is "this" man, and "this" tree.
I am Alfred John Dalrymple.

Nell-

I'm Nyellie Trululu Barton Jangmu Sherpa.

Alf-

I'm a shithead, upon occasion.

Nell-

Sometimes in my dreams…I'm a mindless whore.

From the kitchen comes thunderous laughter.

Aunt Chiri-

Oh, for the God's sake...eat the stew! You can walk
In your sleep tonight! We can hardly wait.

There is light laughter in the dining room.

Alf-

Perhaps we should be more secretive.

Aunt Chiri (in the kitchen)-

No! No! It's all right!

Alf (to Nell…in a lower tone)-

She can't hear this can she?

Nell-

Aunt Chiri has a yak horn hearing aid…and holds it against the wall.

Alf-

She listens to everyone?

Nell-

No! Just to family.

Alf-

Let's not be distracted. I ought to address your concern about being alone, as to the consciousness above man.

Is there an awareness which has to do with you? I speak of Nell Barton…Jangmu.

There have been times when I felt personally touched.

Then…I was not selfless, except that I had been rendered empty, in that I had no thought. A pond… cleared of water…is yet that pond.

I applied no conscious human discrimination to what was before me. As such, I only loved.

One time in the Bering Sea…I expected to lose life.

I came then to love it all, including the bad of it… which part becomes a hateless hate.

Another time, near here, I was moved from hatefulness to emptiness by the act of a child.

Nell-

And you felt acted upon. Touched by the "just there".

Alf-

I seemed connected to an awareness above me, acting upon me, personally. Yes, both times I felt contact…from "just there"…not only as concern about my life, but as arrangement of it.

I seemed to be in a different world, as I became a servant to doing what is there to do…in the hands of a fate that was brought about bymy "loving only".

But…I soon returned to the usual.

Nell-

Intellectually speaking…one time I sat frozen in aloneness. No next instant beckoned. Then you said "Can I help you?"
The incident had no meaning until "this part" arrived.
Of course…I haven't been saved from drowning, or made to love by being cleared.
Sorry I mentioned it!

Alf-

Nell, in this universe tomorrow has not yet occurred. We have "I"…so we make our own choices at the bridge…thus, freewill. And if our "I" is clear and loving only, or nearly so, we are likely to follow "best path". Fate seems subservient to choices of the human "I". Chance enters…and we deal with it as that only. But most of us sense the "just there" part…and it seems to apply to each one, personally.
At those times of being cleared…our actions don't face the need to discriminate or choose, but to do what is there to do, as it is offered.
One is between thoughts. But, being human, we begin to think in our usual way…and soon return to the usual.

Nell-

Al…what is your bottom line?

Alf-

If the soul goes on…the way of contact with the Consciousness "just there" is beyond what we have yet grasped.
Whatever "I" is…as existence…is imperfect neces-

sarily. You and I feel it most clearly when we love.

Nell-

I believe in reincarnation. But I agree with you. Al... it has been improperly defined, this showing in the continued presence of "caste".
As for fate...we haven't arrived at knowing the truth of it.

Alf-

So, Nell...tomorrow, shall we walk up to Pheriche?

Nell-

Yes. Tonight I'll go into the kitchen to be with my aunt and cousins. Tomorrow I'll be your porter.

Alf-

My heart is lighter, Nell. But I need you to carry it.
(shakes his head)
No! I'll carry it! Also...I'll carry yours!

Nell-

I'll carry both! I insist on it!

In the kitchen...pans are being banged.

Aunt Chiri (shouts)-

Will you end the yapping? I need help with tomorrow's stew!

PHERICHE

Pheriche's valley has a handful of buildings...facing the trail running through. To the left, west, is a river... beyond the buildings...showing as a flateness with some water in it. At hand, east, are hills, this remem-bering that the town is about fifteen thousand feet in altitude.

To the right...the hill comes to the buildings, but the steep-ness levels near them. On the hill is a trail...obvious...and sev-eral people are on it.

Alf (points to the right)-
Look, Nell...a new building.

Nell-
It's a hotel...and best for us. The others are smaller
and have dormitory beds. You can get a room here.
Two dollars.

Aunt Kamala runs it.
After dipping to a bridged stream and then getting back to the level, the trail forks...one part continuing north, the other going to the right, toward the new building.
Nell steps to the right.

Alf-
I could fire you. That's because I'm the boss.

Nell (stops and looks at him, with a smile)-
Boss…would you like to stay somewhere else?

Alf-
This one looks suitable.

Nell (walking on)-
My auntie would have been disappointed to know I was here and didn't stop.

Alf-
Tell me about this aunt.

Nell-
Aunt Kamala. She hugs a lot…and you might get a big one.
Also she can be picky. I've seen her tell arrogant asses to get out. Of course, she calls the army men first, so they're present when she gives the order.
The new building seems identical to the one in Tyangboche…and entering it entails going around to the back, to a flight of steps.
As they begin climbing the steps, an older, Sherpa man is coming down.
He smiles broadly at the sight of Nell.

Nell (as she hugs the old man)-
Jangbu. I'm being a porter.
(turns to face Alf)
This is my friend…Alfred John Dalrymple

Alf-
Please call me Al.

Nell-

He's different from other trekkers.
(her face colors…and she clears her throat)
He's smart! I mean…he already knows Rita, from a dozen years ago. And we met in New Hampshire.

The two men clasp hands.

Alf (laughs)-

She thinks I'm smart…because I keep my mouth shut when she gets bossy.

Nell-

He's writing a book…and paying me to listen to his thoughts.

Alf-

It's very helpful. Sometimes she calls me a nitwit.

Jangbu-

Are you writing about here?

Nell-

No, uncle…he's telling me about existence.

Alf-

I don't know much…of course!

Nell-

It's imperfect, he says.

Jangbu-

That surely is true!

Nell-

Consciousness, itself, is the basis of it all, in his opinion.

Jangbu-

It's safe to say that!

Alf-

And so…love is the basis of it all.
Jangbu (smiling)-
Sure!...people will nod their heads in agreement.

Nell-

Your pond must accept the rain…and not be so full of what you already know that nothing else can get in. Love is not love…if it can't alteration abide, as someone said. Probably Shakespeare.
But, uncle, he means there are things beyond our knowledge.

Jangbu-

Yes, boss! And I would agree with him!
(turns to Alf, and smiles)
Al…you and I must speak again, one day. Not that I know anything, either.
I guess we both know about flying near the Sun.

Alf (reaches forward his hand, and they shake again)-

I hope to always keep that in mind.

At the top of the stairs he and Nell turn left, and enter the dining area. The check-in counter is to the right.

Behind the counter an open doorway reveals part of the kitchen.

Nell (shouts)-
It's Nell…come to see you. My husband is away.
(she raises part of the counter, and steps through)
Aunt Kamala (appears from the kitchen)-
Ah, Nell! I'm happy to see you!
(they hug)

Nell (faces Alf)-
This is Alfred. We met in New Hampshire. You can call him Al. We'll climb to Changri La.

Aunt Kamala (looks at Alf)-
Good idea!
(she walks to Alf, and hugs him)
Come…I'll show you to a room.

As she moves through the dining room, they follow.

Nell-
We can talk at that table, the one at the end.

Alf (speaks lowly)-
With the red cloth…as though it's the one we sat at last night. I can see people on the hill.
(studies the hill)
That's because there's no trees…of course.

After Kamala shows them the room, they leave the packs, then return to the table.

Aunt Kamala-

I'll bring the menus. Nell…come to talk with us later. Or, maybe now?

Nell-

I need to listen to Alfred awhile. He's telling me stories.

As Kamala departs, Alf sets papers onto the table.

Alf-

Nell, your old friend we met on the stairs, Jangbu, seems very thoughtful. He spoke of flying too close to the sun.

Nell-

Uncle Jangbu…was telling you not to play God?

Alf-

I think we agreed that a man is flying too high…if he lacks humility.
Nell…over ninety percent of what accounts for the mass of this universe is identifiable only by what it does…as you said earlier.

Nell-

By the way it pulls and pushes distant objects?
Al…our humility is understood.

Alf (pushes aside the papers)-

I'm finished with the essay.

Kamala brings a menu, and they order. Then Nell accompanies her into the kitchen, but soon returns.

Nell-

I ordered hot water for showers. Enough for both of us. Is it all right? You'll need to pay two dollars.

Alf-

Wouldn't it be healthier, up here, to be covered by dust?
But it's all right, Nell.

Nell-

Why is there "Our Something" rather than "Nothing"?

Alf-

I'm finished with writing.

Nell-

No one seems to know what "Nothing" is. Maybe you have a contribution.

Alf-

Nell, I think "nothing there" would be spacelessness. "Our Something" came from so-called "Nothing" only if it was there.
Perhaps "Nothing" is existence nearer to perfection, without need...and that's why it seems to be not there. Only with imperfection, missingness, lack...will energy choose to manifest into that which changes. We see the material part of what effects.
Nell, all man can do is study "Our Something". The

more he knows of what it is…the more he'll know of
what it isn't. Maybe.

Nell-
Put that in your book.

Alf-
I don't know what "Nothing" is.

Nell-
One day…will consciousness die?

Alf-
I assume the laws allowing a singularity to bang…
Were present before ours got to its banging. But…
This universe, or its scattered parts, will die.
I assume. But…who am I to know?

Nell (reaches over…and pushes aside his papers)-
This is between you and I…Al and Nell.

Alf-
That makes me less of a pompous ass?

Nell-
I know you're not that! You're only a man…but a
good one…sincere and kind.
It makes sense that this universe will one day die.

Alf-
Nell, you asked about a larger mystery. The entirety.
Will Consciousness die?
When Consciousness suddenly is, the love of itself

and its continuance suddenly is.
Love, or need, keyed imperfection as a necessity within the going onward…and the wanting to.
Without love's imperfection there would be no allower of mass, laws, movement.
Energy equals mass…because Consciousness effected and effects.
If "Our Something" loves, "Our Something" is.

Nell-
In one reality…or countless of them?
Al…what about our souls? Will they die?

Alf (sighs)-
Who knows? But…you speak of realities as being linear. Does every carpet have a final "different from"? Does each part of it take its sameness to the edge of doom?
Nell…how many billion years do you need?

Nell-
Can anything be perfect…and have meaning?
Speaking of heaven…if "Something" doesn't have lack, how can love be there?

Alf-
Robert Browning said that a man's reach should exceed his grasp, or what's a heaven for?

Nell-
You're saying you believe in heaven?

Alf-

I didn't say that! I'm like many humans…who have no idea what "Our Something" came from or where it's going.
I'm trying to, but…
And, Nell…there are other dimensions not only beyond our knowledge but, also, our imagination.
(he holds his head)
If so-called "Nothing" is there as an entity…it would have movement. But why would it? One time I dreamed…"Nothing" is salvo lacking lack".
(he gets up)
Movement of what? In my logic…perfection has no need to move, nor does it have a "why" or "why not?"

Nell-

Salvo…? Doesn't that suggest a beat…bang, bang, bang? Or…a rhythmic movement?

Alf-

Aren't we to close to the sun, Nell?
I'm finished with this!
(he turns toward the door to the rooms)

Nell-

Where are you going? Al, I'm sorry! I didn't mean to upset you!
Please sit!

Alf-

I'll come back.

Nell-

The sun is gone! Light a candle!

Kamala arrives with their food.

Kamala (puts her hand on Alf's shoulder)-

Please...sit!

Alf (after he sits, and the food is allotted)-

Let's let it go, Nell.

There is silence for a minute or two.

Nell-

Will you write your book?

Alf-

Yes...I will. It'll be a quick essay, thrown into darkness.
(he smiles)
Or...I could make it a conversational piece, by putting you in it. That would bring back the sun.

Nell-

Oh, do you think that? After I've been such a pain?

Alf-

I could make it a dramatic essay...a sad one.

Nell-

How? Why?

Alf-

Assume we're standing at a cliff's edge. A man delivers a note from your father. Well, two notes…one to you and one to me.
Mine says "I made a female robot, and named her Nell".

Nell-

What a tragic thought! Al…what would you do?

Alf-

I'd throw you off the cliff!

Nell-

Why?

Alf-

In the essay, I'm in love with you. So…I throw you off.
I can't stand the pain of suddenly knowing you don't exist…Nyellie Trululu Barton Jangmu Sherpa.
Over you go!

Nell-

What did the other note say? The one to me. You had It in your hand.

Alf-

I look at it. It says: "I've got this Nell with me. Wish I could've made it more like you, Nell. I'll change this machine's name to Priscilla…I always intended to.

Nell-

Oh! That is sad. What would you do?

Alf-

I'd jump off!

Nell-

How sweet!

Alf-

When I hit the bottom…I'd land on you.

Nell-

Our souls wouldn't feel it.

Alf-

Then we'd be together sixteen billion years. It's all right!It would be non-zero time.
But…when I jumped, I'd end the essay.
Describing hitting you at the bottom wouldn't work.
The reader would assume I'm dead before that.

Nell-

Not if they believe in an afterlife.

Alf-

Why would they want to know I bounced fifteen Feet?

Alf-

So did I…bounce that high! It's very romantic.

They eat dinner slowly, with few words. Trekkers nearby are speaking of climbing Kala Patar, a foothill of Pumo Ri, and from its crest looking down at Everest's base camp. After eating, he takes Nell's hand and they enter the dark corridor.

Alf-

I don't have my flashlight. Do you have a candle,
Nell?

Nell-

No.
(grabs part of his jacket)
It's the fourth door on your right. Just feel along the
wall.

Alf (as theymove forward)-

I'm at a door. All right...I'm past it. Coming to
Another. Ouch! I hit a nail.

When they arrive at their door, he needs to get the key to it from his pants.

Alf-

My right hand feels bloody, so I'm reaching with my
left hand...to my right pocket.
(he feels Nell touch his right hand)
Don't! You'll get bloody, too.
Here's the key.

Inside, they get a candle lit, and he finds the first-aid kit in the top of his pack.

Nell-

I've got paper towels. Sit on the bed, near the light.

At the edge of the table near the head of his bunk, she melts wax and fastens the candle. Then she dries his hand, and puts antiseptic ointment and a bandaid on the cut. He slips off his shoes and, relaxing onto the bunk, pulls the sleeping bag over him.

Alf-

I'll rest here for a minute. Will you unzip my bag?

Nell (moves the zipper down, then gets on his bunk)-

Let me get it over us, all together. But, remember, our water will soon be hot.
(she rests her head on his chest)

He puts his arm around her, his hand touching her head, and holds her against him. With his left hand he pulls the bag over them.

Alf-

Why are you snuggling with me, Nell?

Nell-

It's cold! Twenty degrees, maybe. Why do you have your arm around me?

Alf-

I have to say there are things about you...I love.

Nell-

Is that like saying "I don't love you!"?

Alf-

It's like…not saying a certain thing.

Nell-

Yeh! There are some things about you I like.

Alf-

That's like having two locks on your door.

Nell-

You said "Fortune plays queen to a man's courage".

Alf-

Is that true now, Nell…in your opinion?

Nell-

Don't know!

Alf-

Shall I say I'm afraid to open my door? I'll say it..if you'll admit to some fear of your own.

Nell-

It scares me to think about telling someone I love them…and being put outside the door.
(snuggles against him, more closely)
It's warm in here. Do I need to leave?)

Alf-

Only if you need to.

There is a banging on the room's door.

Aunt Kamala-

You need to go down the stairs. Your water is ready.
I made enough for two.
Stop in the kitchen for towel and soap.

Nell (with her head up...shouts)-

We'll be there!
(again rests on his chest)
Al...to get the water heated she burned quite a lot of wood.

Alf (pushes away the bag cover)-

Let's do it!
Aunt Kamala (yet outside the door)-
After you get there...and you're ready...bang hard on the pipe.
Bring clean underwear.

In a few minutes they're through the dining area, and to the stairs outside. He goes first, flashlight in hand, down to the flat-rock patio and the shower building.
Inside, the walls are lined with tin sheets, and there is a bench and table. On the head-high shelf is a can with a candle in it. Wall pegs are at hand for clothes.
In the middle of the room is a hanging pipe fitted with a shower nozzle.

Nell-

Light the candle, Al. There's one on that shelf. Hang your clothes on these pegs. Just wear your shorts.
I never did this before.

Alf (lights the candle, then begins to undress)-
You never did what?

Nell-
Have a shower with a man. We've got one minute before we need to bang.

Alf-
How much water will come?

Nell-
Enough so we can get wet and use our soap. There's some little bars on the bench. Extra.
I'm ready.
(she stands under the pipe, clad in underwear)
Then…in a bit, they'll pour enough for us to rinse.

Alf (stands facing her, and is clad in shorts)-
How can you be so cold at a time like this?

Nell-
It's cold! Get closer to me, then bang the pipe.

Alf (puts his hands on her shoulders, and pulls her to him)-
My chest is hard against your breasts.

Nell-
That's not the only hard thing. Control yourself!
Bang it!
(looks at her watch)
Good! Now wait! We have twenty seconds. Oh…I need to take off my watch.

(runs to the table and back)
All right, Al. Bang it!

The water comes in a spray less intense than he expected, but it is hot.

Alf-
God! That feels great!

Nell-
Shut up, and apply the soap.
(stands against him)
Get my back!

Alf (reaches around and applies soap to her back)-
Get mine, Nell!

Nell (gets his back)-
God!...you're hard. I mean your muscles. Stop it!
Wash your hair! I'll get mine now.
Put an end to that!

Suddenly the water is off. Alf works soap into his hair, then he grabs her shoulders, and pulls her to him.

Nell-
We don't have time for romance!

Alf-
Too late!
(he grabs her butt cheeks, and holds her…off the
Floor)

Now there is a voice from above.

Aunt Kamala-

When you're ready…just bang!

Nell-

Do you love me?

Aunt Kamala-

Why do I need to love you…to pour water? Of course I love you.

Alf-

Yes, Nell…I love you.

Aunt Kamala-

My mistake! When you're ready just give it a bang. No great rush…you're probably filthy-dirty. I wish we could see…to advise you properly. Use the table, for the God's sake! Anyway…bye, bye…and we'll wait for the bang. Make it a good one!

He carries Nell to the table. There follows several minutes of hot, slippery, pounding oneness.
A knock on the door occurs.

That door is locked…but not the one on the table.

Voice at the door-

We're waiting for a shower, too.

Aunt Kamala's voice comes from above.

Aunt Kamala-
Are you ready? It's time for me to pour your rinse.

Nell (shouts)-
That's it! Yes! Stop it! Stop!

For awhile there is thunderous silence.

Aunt Kamala-
It has to come…soon! Are you ready?

More silence…for almost a minute.

Nell-
Yes!

Water comes through the pipe.

Nell-
Stop! Don't stop! Oh, don't! Yes…do it! Stop!

The water stops. Another minute passes.

Voice from outside the door-
Are you almost finished?

Alf (shouts)-
Almost!
(whispers to Nell)
Let's get under the pipe. I'll tap on it again.
(carries her to the pipe)

Nell-

Can we pretend we're the German couple?

Alf-

No! She's your auntie. We'll tell her the truth.

Nell-

There'll be hellish anger…but I'm with you on it.

He bangs on the pipe.

Aunt Kamala-

Is this Herr Spangenberg?

Alf-

This is Alf and Nell.

From above…for a moment there is silence, then a pounding of feet, in salvo. This is followed by more silence.

Nell-

Al…she was stomping on the floor.
Aunt Kamala (louder)-
Are you ready?

Alf-

We're ready, Aunt Kamala…truly!

Aunt Kamala-

Alfred…are you ready?

Alf-

Ready as I'll ever be!

The hot water comes again, and they get rinsed.

Voice from outside the door-
How much longer?

Alf (yells)-
Just one minute!

As they dress...there is shouting outside.

Aunt Kamala (to the Germans)-
The hot water is gone. Also...our wood for the fire is depleted. We are sorry! Can you wait until the return to Namche?

German man-
They should be punished!

Aunt Kamala-
They were just married. We must understand what that does to the soul.
(she laughs)
Good things!

German man-
In Lukla...they gave us yak meat so tough we made them take it back to the kitchen.

Aunt Kamala-
Then the lady in the kitchen beat it with a hammer. She brought it to you, and put it firmly onto the table, and waited for you to complain. But you wisely kept your mouth shut.

German lady-
Shouldn't we be angry that the shower didn't come to us…that you promised?

Aunt Kamala-
Yes! Be angry! Then…understand!

Alf and Nell are dressed.

Alf-
Let's go, Nell. We'll face the music.
(they walk out)

Aunt Kamala-
There they are. Look at them!

Alf and Nell stop near the German couple, who are of middle-age. The husband is heavy, and the wife is trim.
The woman goes to Nell and hugs her. The man opens his arms and smiles at Alf, who steps forward and gets a hug.

German woman-
Congratulations to you both!

German man-
We didn't know the truth.

At the top of the stairs the old man, Jangbu, appears, and studies things.

Nell (to Jangbu)-
Uncle, were you in the kitchen…through it all?

Jangbu-

After Kamala jumped up and down…I pointed to the water…as though to say "Be kind to them…they're out of their heads!"

It would be good for more of us…to be out of our heads.

Alf (to Jangbu)-

Thank you…my friend.

Jangbu (laughs)-

Now that you're married…you won't need to talk so much!

AFTERWORD

On the Everest trail...Alfred hires Nell Barton Jangmu Sherpa to be porter, and to listen when he reads from his new book.

Alfred believes that the Ground of Being is conscious-ness. He thinks time, distance, and size are part of it, as is mass and its laws.

He says Consciousness effects...this as to movement and immediacy...allowing energy to be mass. Energy equals mass. E=M.

The immediacy of Consciousness allows reality to be separable...assuring freewill.

The human "I" can make choices as separate from genes, other men, or the "I" of the Ground.

Nell tells him his mother's milk had laughing gas in it.

She questions, criticizes, often agrees.

Then, very near the sun, they shower together and wash away the dust of human thought.

CONSCIOUSNESS AS THE PRIMARY FIELD

Fate and Freewill

Alfred John Dalrymple

THE CONSCIOUSNESS FIELD

Part 1

Possibility
Alf and Nell are in the Aleutian Islands. They sit in a lone cab-in...at a window table. There is evident, at the periphery of the moment, an inlet-like body of water touched by treeless hills.

Alf (taps a pile of papers)-
Nell, this is my last essay about Consciousness.
right or wrong...this is my say.
Why do you look through the window?

Nell-
If our boat drifts away, we walk to town. How long would that take, Al?

Alf-
Over the hills...about five hours. Are you in the mood to listen? Tell me the truth, sweetheart.

Nell (reaches into a packsack)-
Oh, here's a donut...to go with my coffee. I'm ready!
(she breaks the donut and gives him half)
Are we getting married...next Sunday?

Then we go to Nepal and have a ceremony with my Sherpa family.
(pours coffee from her cup into his)
I'm eating my donut.

Alf-

In awhile I'll have mine.
Nell…would you agree that every existent was preceded by the possibility it could exist?
I mean…after Consciousness continued as "Our Something"?

Nell-

I'll say "Yes!"…and that I think existence didn't begin with this universe. Is that your point?

Alf-

I call the larger reality "Our Something". The entirety of it is a field…the Consciousness Field. It moves manifest parts of itself onward…through encounter.

Nell-

Are you making a comparison to the Higgs Field, and its particle, a boson…as it gives mass to what it encounters in this universe?

Alf-

In a way. I'm saying Consciousness is the primary field. But…when it encounters, it may not need a particle, Nell. And…because it is an immediacy, the action of the encounter can begin from outside or inside. Also…the action is constant in an ongoing manner.

Nell-

This immediacy is key, to you, isn't it, Al? It allows you to defend cause and effect.

Alf-

Nell, the entirety is carpet-like. Changes in it are of connections...but throughout, an immediacy prevails. And...

Nell-

We go on the "same as" and "different from"...you said. Remind the reader of the separable nature of things.

Alf-

I will, Nell.
(reads)
Within Consciousness' all-encompassing condition are universes...and they present time to us in a slower way, as one thing following another, relative to the speed of light.

Nell-

Al, let's go back a minute. You say we're preceded by the possibility we could exist.
How could any oneness be the first thing?
Did it enter into something already there...and change it to "Our Something"?

Alf-

That seems likely. "Our Something" is imperfect necessarily...because the only allower of continuance is love. I think all things were born and continue to

be…through lack acknowledged.
"Possibility" entered with the only oneness which is all things…Consciousness.
To manifest…twoness was effected, for needed laws. Consciousness "becomes"…as each part continues the "same as" and, so, connected…but also "different from" and, so, separate.
(he reads)
"Possibility" necessitated imperfection…which is central to the definition of love. There is need, lack, missingness.
As for perfection…it lacks lack. So…screw it!

Nell-

Unless it's perfect imperfection. Oh…Al, is that what "Our Something" is?
Nah!How can I say that…and watch a child die?

Alf-

It's not easy to love the bad as well as the good. Love has to do with lack, but man's choices often move toward perfect crap.
(he reads)
In non-zero temperature, Consciousness enabled continuance of itself, through change.
Consciousness…or "I"…engendered "best path" for manifest existents. So we have "twoness" allowing entrance of laws.

Nell-

Al, at the beginning of "Our Something", if the temperature was at least non-zero, aren't you saying there was an "into"? Something had to be present

for any temperature to be…even if it was almost none.
This field you speak of…effected itself forward changing what it encountered.
But…you make it seem as though Consciousness was the only thing that went forward.

Alf-
All things in "Our Something" exist as an enablement Of Consciousness. They're part of it.

Nell-
Al…what are the known forces?

Alf-
Electromagnetism, the strong and weak nuclear forces, and what gravity does.

Nell-
So…Al…physicists say the Higgs Field and its particle is throughout this universe…and that it effects mass to the extent we wouldn't be as we are without it.
You're saying…these forces, and all else, were enabled by the Consciousness Field.

Alf-
Yes. "Our Something" arose when love entered, as acknowledged lack.
Consciousness enabled "best path" in every unit of change.
Each unit changes and continues.
Nell…things took on meaning and purpose when

enabled to go forward wanting to do so.
The "manifest" goes forward as each quanta unit being the "same as" and "different from" itself.
Each unit has three parts...the past part of it, the "now" part of it, and the future part.
The "now" is the ethereality of it.
Remember, Nell, Consciousness is immediate.
When a unit is of "encounter"...the immediacy in the "now" enables a going back and ahead in time.
In our universe, when an electron meets its anti-particle, the positron, they couple and annihilate. In action faster than the speed of light. It was suggested by Richard Feynman, in support of Einstein's thought, that this causes a move back in time.
The result of the annihilation is the emitting of some photons.
I used this comparison to draw attention to what happens when the speed of light is exceeded.
As Consciousness "encounters"...the connection in the "now" part, allows action not of annihilation but of enablement forward.

Nell-

With movement back and ahead...connections are allowed?

Alf-

Nell...what is it that allows you and I to exist?
Change! If we stayed the "same as"...we would fade away or, rather, we would return to what "Our Something" entered into.
Each instant...we become "different from", but we Retain the "same as" forward. Both must continue.

Only connection between the past and future, in the sense of the "same as" and "different from" allows meaningful and purposeful continuance.

Nell-

As I listen…I agree that movement of particulate things can only happen at the smallest units of existence. You can't leave anything behind.

Alf-

That seems true. Concerning continuance of existence, whatever moves forward accompanies action at the smallest parts.
Also…reality moves forward not every minute but every instant.

Nell-

So…Al…if particles are enabled forward at the immediacy of Consciousness…there's no death, other than the "same as" going on…also…as "different from".
The "same as"…in the alteration…never dies.
"Death once dead, there's no more dying then".
Who said that? Oh, yes, it was Shakespeare, in that sonnet of hers.

Alf-

Could be, Nell. And, surely, she was eating a donut.
Just kidding!
I need to say how much I appreciate you listening to me.

Nell (puts her hand on his)-

Al, particles in the Consciousness Field move forward as each "now" has "encounter" faster than the speed of light.
Connection occurs between the past and the future.
How far?

Alf-

Far enough for connection to move units forward the "same as" and "different from".
And I do apologize for repeating myself.

Nell-

How will we ever know what is the smallest part to be encountered?

Alf-

I can only say…in my ignorance…that "now" encounters to the smallest part. And…each unit's "now" moves forward according to allowance by action which is of immediacy.
The past and future of each unit…connect.

Nell-

Al…don't lose your confidence.

Alf-

I'm only a philosopher. And I have great doubts. Who am I to think I know about these deeper things?

Nell-

I love you. Buck up!
You say reality is carpeted.

Alf-

What "it" wasn't, time microfibers.
Sorry, Nell…one night that popped in.

Nell-

Al, I'll adopt the baby…and let it suck. Did I tell you, I'm a reincarnation of Shakespeare. Nah! But Nyellie Trululu ran through the heather…mindlessly, upon occasion.
Get on with your book.

Alf-

Nell, I need to discuss "entanglement", where two bits of matter, apart by vast distance, co-respond when stimulus is applied to one. If referring this to our slow time, no message could have been sent between the bits, because it couldn't exceed the speed of light.
This thinking kills acceptance of cause and effect, and freewill. Some want to say "Goodbye" to those truths…and "Hello" to man as robot.

Nell-

You say, Al…that by the immediacy of Consciousness, message could have been sent.
I must get to town…I need chicken. Not really!
So…you're agreeing that nature is separable.

Alf-

Yes! Cause and effect stands. Man has freewill. You and I are not machines…we're human.
I need to talk about the "now" part.

Nell-

Get to it!...my man.

Alf-

During movement forward, "now" has action faster than the speed of light, for each connection made as to contact within manifest parts.

Nell-

Alf…no connection, no future? Consider this: it's what happens when you almost annihilate my donut.
Or not.
Sorry!

Alf-

Good point!

Nell-

Al, the "now" part must connect ahead, as though all is written?

Alf-

Nell, in parts with pure consciousness only, "now" is ethereal, but it doesn't have "I". It has "best path" as to Consciousness' continuance.
Consciousness is choosing for it, by letting it be itself as to going onward.
So…Nell…let's bring the particulate and ethereal down to you and I and this table. Nah! For now we can leave out the table.
Our body and body-mind, with the atoms and their parts…down and down, to the quanta units…have

pure consciousness. They self-maintain...as "now" occurs in each, and there is accordance with laws.

Nell-

That which is manifest follows "best path".

Alf-

Yes. We shouldn't blame our body and body-mind, when on their own, for acts of self-maintenance that seem short-sighted. They have an ethereal "now" part in the movement forward, but they don't have "I". Thus the wildfire burns.

Nell-

Al...that's your duality, you once said. The body and body-mind is this part, and the other part is "I".

Alf-

At choice's bridge the "I" chooses separately from all else. I mean...it can.
And...between thoughts, or at least when free of the worst parts of human discrimination, our "I" can be somewhat pathed to the "I" of it all.

Nell-

Is that "I" sometimes "just there"...as though ten feet away?

Alf-

I once told you of my experience in that regard. Do you remember?

Nell-

Yes, Al. You were above Lukla, and returning from Everest. You were in the condition of "I hate you morons!", which will occasionally catch those who have been at altitude.

Alf (laughs)-

I was full of the "worst part". If you said "Hello", in passing, but also picked up a stone, I might have guessed you were intending to hit me in the back of the head and steal my money.

Nell-

Then…very near a small Buddhist Shrine, a little girl gave you flowers. She smiled, and walked away. She asked for nothing. It was merely an act of kindness.

Alf-

Ah, Nell…the act rendered me empty.
(he sighs)
As I walked on, an English lady, passing, asked: "Do you have a wooly cap I can buy?"
I said "Yes…at the top of my pack. This morning I took it from the bottom." Then I added "It's new…I bought it in Kathmandu and never did wear it."
Of course, the two happenings seemed joined, and connected to something "just there".
As I walked on, toward Lukla…I was, thoughtlessly, freed from what I call the "worst part".
Nell…I only loved.
In awhile I knew I could remain in that condition,

or return to the usual…which I soon chose to do.
Surely, Nell, you have felt a presence.

Nell-

Yes, once or twice, in sort of a "feeling only" way.
Al…as you said, and I agreed, sometimes we all "love only", in the sense of forgetting ourselves. We get "outside" that way. You usually call it being between thoughts. And…often…we are rendered empty when encountering what is beautiful.
But does anything out there, or "just there", care about Nell Barton Jangmu Sherpa?

Alf-

Nell, the clearing of the mind is available and, when it occurs, something can seem to be "just there".
You can always wonder about what caused the happening. Some would call it an arrived feeling within a fate brought by the choices you made. Some might call it destiny, saying you had nothing to do with the "writing" of it.
No matter how it's defined…you will feel personally touched.
I need to move on, sweetheart.

Nell-

Some would say it's like a clock. God wound it up by making the conditions…and now there is a "hands off" settlement.
As you say, Al…we are alone in our choices.

Alf-

Sure…but that doesn't stop something from being

"just there".
(he reads)
Entire "I" is the enabler of "Our Something". In our containment, those possessedof "I", such as Alf and Nell…when they are between thoughts and, therefore, "loving only"…can have an opened path concerning contact with Entire "I".

Nell-
Al, most humans assume divinity, and call the enabler…God.

Alf-
Nell…sure.
(he reads)
Our "I" has immediacy, which means it directly partakes of reality's zero time and distance. It's not hampered by a particle's relation to the speed of light. "I" is not limited by here "then" there, but it is here "and" there.
At the bridge, "I" is free to look ahead and back, in the way of choice. This can separate him from DNA, and other men's directives.
As for being separate from the enabler of "Our reality", the desire of the enabler is for those with "I" to be of humanity not of robotism. This in the choice, not in the loving.

Nell-
Al…sure…we must accept that on the carpet we are responsible for the connections we make.
But…it is time, in your essay, to move past that.
I agree that the human "I" is ethereal. But how can

it exist outside the body? I'm speaking for those who will read your book. Also…I'm speaking for me…a Sherpa girl you found near Everest. I'm a University of New Hampshire graduate, but my ancestors accepted reports of monks occasionally leaving the body.
I believe it, but how many of your readers will?

Alf-

Nell…we agree. Although the human "I" feels tied to the body, it can depart.

Nell-

Some of my thin-air ancestors worked hard to be able to.

Alf-

Probably after awhile they were not so fearful of leaving a possession which seemed desired but not entirely needed.
Nell…when I was thirteen, one day I left my body. This occurred in a school classroom, as I was reading aloud. Suddenly I was above my body by about six feet. And…as I observed the scene, my body and body-mind, below, continued to read aloud.
I was present in the "up here", the same way I'm here now. I seemed material, and with my "I" up here. Am I here, Nell? I have no doubt that I am.
"I think, therefore I am" as a Cartesian proof of existence, assumes I'm where I accept that I am, according to feeling and reasoning.

Nell-

Also, Al, if I'm here…and accept that I am…you are here. But of course that verification needs to have at least two people on Earth.

Alf-

In the classroom my "I" was above, observing, as my body and body-mind continued to read aloud. This sort of departure is usually reported by those desperate to get away. For instance, it's not uncommon that a mountaineer who is facing peril will leave his body, and observe…as it continues the climb.
When I was above, in the classroom, I wondered "Why can't they see me?"

Nell-

As you know, Al…some call it projection.
Those who have the experience don't call it that… this group including humans who have departed while undergoing surgery, or mountaineers and others arrived at an outer limit.
Centuries of Buddhist monks, and such, did it for other reasons, without fear causing it.
Al, reason is expected to explain where the "I" is, but I think consciousness' grasp of its location depends on feeling, as well. If I kick my toe, it hurts… here…without any need to think about it.

Alf-

I agree, Nell. I didn't kick my toe as I was "up here", but my experience, which included "Why can't they see me?" had strong feeling, at least in the sense of "wonderment".

Nell-

When you were above...did you feel possessed of all your faculties?

Alf-

I think so. I know how a thought, in its ethereality, relates to feeling. I had wonderment. I love you. It's the "I" defining the difference between those acknowledgments and the action of a wildfire whose results connect the past to the future. You choose to love. The wildfire is a sharer of Consciousness' Continuance.

Nell...in thinking we're superior, we often eliminate what we think is beneath us, or always harmful.

Love is the basis of continuing existence, and one of the ingredients is wildfires.

But, yes, I seemed to have my usual faculties.

Mostly I was amazed. It seemed that I was in a different dimension...and yet I was in the same place as everyone else in the classroom.

Nell-

Why did you return to your body?

Alf-

As you know, Nell...we love our bodies. Also, I was anxious...afraid of this unknown condition.

Once well beyond the happening...I wondered about the difference in the "up here" place. I got a feeling for ethereality that supports the experience of it in relation to the body and body-mind, as to how "I" needs the body and loves it, but can separate entirely from it, and continue, itself, "entire".

Let's move on, Nell.
(he reads)
"Entanglement"...where two bits of...

Nell-
Whoa, Al...you already discussed that.

Alf-
You're right! Sorry!

Nell-
Al...is "entanglement" what Einstein called "spooky action at a distance"?

Alf-
Yes, Nell.
(he reads)
As to Consciousness being all existence, the field is ethereal and manifest, as to being particulate.
Manifest parts self-maintain forward...through the particle-like encounter. With deeper thought we can apply this action to cause and effect, the separable nature of things, freewill, and fate.
Remember...consciousness has immediacy.

Nell-
Put that on your back cover, Al.

Alf (reads)
All parts of manifest existence have "now". It is Consciousness continuing and, as such, sharing...with action of self-maintenance, thus "best path".
When "I" is departed, the body and body-mind, are

not robotic, but maintain relative to memory, and to "best path".

Nell-

Then...Al...we are entirely human.

Alf-

I think that's true, Nell.

Nell-

Al, you're telling me...that when Consciousness is manifest...we ought to acknowledge the enabler as a field which effects the "manifest" forward. I can see physicists accepting this. But I think most others will be slow to do so.
There is great power in the "just there" part of reality. A felt presence would be thought of as manifested by a being...God. For most persons, Consciousness, as "I", didn't make the "just there".

Alf-

I think you're right, Nell.

Nell-

And yet...certain monks in Tibet were capable, it is believed, of having their "I" depart and travel to another location...let's say Lhasa to Shigatze...and during the occurrence be physically seen in both places.

Alf-

I've heard of such action. We are expected to accept that the "I" of the monk, the ethereality of him, could enable manifestation of his body…elsewhere.

Nell-

What do you think, Al?

Alf-

I think it depends on the power of that "I". Is the monk fully connected, as to himself and the "just there"? Is the path entirely clear…or nearly so?
Nell…how many have been in that condition? Jesus, Mohammed, Buddha. A few others have achieved "enlightenment".What to say? A few?
Could the monks have been cleared to the extent they were able to depart their bodies, travel, and manifest themselves at the other location?
The answer is "Yes"…if they did it.
But…for a human to do that, wouldn't they need, first, to spend many years meditating…as did the monks? Or at least be cleared for a lengthy time… which Is difficult for a man to achieve.
And then…how many could depart the body and travel more than six feet?
The traveling part is one consideration, the being able to manifest is another. I would guess those who who could do both were very few.
Ah, Nell…who am I to know?

Nell-

Al…they spoke of one or two.
I believe at least one monk was able to do that. He

was seen in two places...Lhasa and Shigatze.
I'm a Sherpa girl...who was a child in the region of Khumbu, at Everest. There is something in me that wants to believe the monks could manifest their bodies.

Alf-

Maybe they could, Nell. You have a right to your beliefs...which are the same as mine, in what we think is possible.
(he reads)
Time and distance is Consciousness "becoming".
Even with "change"...Consciousness is immediate.

Nell-

Connected and separate? Yet...as a oneness.

Alf-

Yes...of the manifest, the "different from" allows Nature to be separable.
Cause and effect depends on sameness and difference continuing onward, as to time and distance...in our familiar linearity and, also, in a carpet-like "becoming" oneness.

Nell-

In a continuing "now" kind of way?

Alf-

So...briefly I'll return to "entanglement".
(he reads)
Due to Consciousness' immediacy, when two bits of matter co-respond at vast distance, there was

message sent.
It has to be so, Nell…according to our humanity. If cause and effect stands, nature is separable, and man has freewill. He can make choices separate from God, DNA, and other men's directives.
Man is not a robot…he is human.
(he sighs)

Nell-

Do you sigh because you stand at the edge of an Abyss, and look down at Robot City?

Alf-

Even those things without "I" move according to "best path"…this helped onward by the laws which Have been established.
This is true of the parts of a robot…its atoms. As for the robot…only the atoms are true to "best path". The assembling of the parts is manipulative as to the service not of itself but of another entity. The "program" goes on until the death of the program. The atoms go forward as atoms, unto themselves…that is, as iron or copper or such.
Ah…crap!
(pushes aside his papers)
Soon we won't know what we made. Can this "thing" think for itself? So…is it a thing, or did we make a human?

Nell-

Don't be discouraged, Al.
(she reaches to his pile of papers)
What are these other three or four pages?

Alf-

A few words about the dangers entering with some of the new technology. For instance…how we think via computers, and cell phones. I see those things weakening our minds…maybe cutting down on the use of it. I…

But…Nell…most will ignore criticism of the new.

Nell-

That's why you need to speak of it…for the few who listen.

Alf-

Man doesn't feel responsible for results of the use of most new things. He sees the arrival of them as being written. Results concerning their use…are fated onward.

Also, blind service is in the mix. If it is suggested that making robots could hurry the death of man's self-reliance, mainly as to thinking for himself, how many would curtail the making of robots?

Nell-

Al, man often seems blind concerning the obvious. When we see pictures of imagined humans from the future, they are slim, of medium height, and smart. Isn't it clear, today, that most humans are going forward in a way different from their picture?

Alf-

Nell…soon when you pick up a phone…you'll only need to say "Call Frank!"

Nell-

They already have those...and they're wanting to imbed them under the skin of your wrist.

Alf-

Nell...dance for me!

She stands...and begins to dance.

Alf-

Take it off!...the crazy man yelled.

Nell (gyrating)-

Tell the crazy man to put wood in the stove.
(when he steps toward her, she laughs)
No! That stove!
(she sits)

Alf-

But...why not in both?

Nell-

I'd be good for nothing the rest of the day.
(she reaches for her thermos, and his cup)
Have coffee...and get on with it!
What is love? Explain the open door policy.

Alf (reads)-

When lack and need occurred...consciousness, as love, necessarily included imperfection.
Reality goes on in you needing to be up if you're down...in your positive needing a negative for its definition...and...Love...?

Nell-

Is lovely imperfection, Al. I need you to massage the back of my neck.
(he does)
People who get massages live longer, they say.
You can continue.

Alf-

With what…the rest of your body?

Nell-

No! With your paper!

Alf (shrugs)-

Consider this, Nell. Why do all things, even those within seeming chaos, maintain themselves? This table does, because atoms do, because quarks do, because…on and on…down and down.
This said not referring to the far journey, but to maintenance along the way.
Atoms and quarks maintain as to "best path". They don't err…as does man.

Nell-

But…Al…you want that freedom. As you say, we'd be machines without it.
We learn from those times of adversity.

Alf (reads)-

"I" is part of an ongoing reality. When you arrive at what allows an "open path", and have contact with the "I" which is "just there"…you will, soon, return to the so-called usual.

As Buddha sat before the tree, and became one with it, neither was "arrived"...but as Buddha went back to the usual, he was different from what he was before the encounter.
When you and I return, our pond ought to be less full of ourselves, as to attendance...this providing more room for the rain.

Nell-

By rain, you mean the truth of things? We should let things speak for themselves.

Alf-

What is "there"...before you...must be as it is, to the greatest extent of that, and not to the greatest extent of your wish.

Nell-

Hard to do, Al. We almost continually need to discriminate between things.

Alf-

As to our developing character...we relate to an ideal ethic, as to what we think we ought to love, but also to what we love without thought.
We have emotions that are appropriate...such as anger, desire, joy, happiness, sadness. And all these can be used in the wrong way.
We err...usually on the side of poor judgment, rather than viciousness. For instance...anger and sadness are, clearly, often appropriate, and then we keep them around too long.
What were we talking about...before?

Nell-

The electron in an atom maintains as to the laws,
in a "best path" way. It doesn't make errors.
A man maintains according to the laws...and...
Al, it's a big "and" to stand before, alone.

Alf-

You use your head to have proper ethics...and you
clear it to let in what can slake reason's thirst,
within reason's blindness.
Then...being both full and empty, you do what is
there to do.
(he holds his hand toward her)

Nell (takes his hand)-

Then you say "I love you!"

Alf-

You yell "I'm free! I am not a robot! I am..."

Nell-

A nitwit...did I forget "nincompoop"?
And...you won't say you love me!

Alf-

Why wouldn't I say it?

Nell-

It won't roll off your tongue.

Alf-

Why not? It's not stuck there.
I luhluhluh…I luhluhluh. Oh! It won't come off!
I…love you. Ah! It rolled off.

Nell-

Are you finished with the essay, nincompoop?

Alf-

Yes, Nell. I could write another part, a third one,
asking this: "When I choose to move my right arm,
and it moves…how did my "I" touch my body?
But I'd be drifting.

Nell-

So…you won't write about it?

Alf-

Those without "I" share "best path" as encounter
occurs, but when "I" is having to do with, personally,
there is a radiated effect we don't understand.
How do I touch my arm…by thought?
Is particle to particle involved?

Nell-

Al…the Consciousness Field enables manifest things
to continue onward. Perhaps the action is particle to
particle. You can say it is.

Alf-

The action seems to be that…but I think the truth
will be hidden awhile, Nell. The mystery of the
contact lies in the "immediacy" of it. In each unit of

existence…is the "change" enabled from out, or in… or both?
My human "I" connects to the manifest…when my arm moves. What else can I say?

Nell-
Leave it, Al.

Alf-
I don't understand Consciousness.

Nell-
But don't be discouraged!
You only need to be sincere.
Al…do I believe every existent is preceded by the possibility it could exist? My answer is "Yes!"
There's your ethereal base, Al.
Stick with it! Let it roll off your tongue.

Alf-
I love you, Nell. See how easily it rolled off?

Nell-
You love your body. Do you love my body?

Alf (reads)-
If "twoness" was enabled…in non-zero temperature, where there seems to be nearly free flow of what can radiate…continuation means "positive" has its "negative", in the same way "up" has its "down".
I'm done, Nell.
(he pushes away his papers)

Nell (comes around the table to him)-
I'm part of your audience. Do I listen?

Alf (slides his chair away from the table)-
Sit on my lap, sweetheart!
Is there a dog called the Lhasa Apso?

Nell (sits)-
I sit on your lap, and you think of a dog?

Alf-
When I was little, we had a dog named Bobo.

Nell-
I had a worm named Frank. Al…so what…?

Alf-
I don't know! What did Frank look like?

Nell-
What about Bobo?

Alf-
A German shepherd…sort of…but tan, and his tail went up. He was medium in size. He didn't start fights, Nell…but, once in…he never lost one. Except… ahh…!

Nell-
What is it, Al? What happened?

Alf-

In Portsmouth, New Hampshire…when I was five or six, and looking from our house's window, into the gated yard…there was Bobo, asleep. Then, a huge Saint Bernard leaped over the gate. I knew him as one Bobo had recently whipped.
I need to repeat, Nell…Bobo was asleep.

Nell-

He attacked Bobo while he slept?

Alf-

I ran quickly as I could…out to that scene.
Nell…it was a long run. I must have gone through the house to the back door, then along the side of the building.
No! That's not the way it happened. I used the front door, not far from the window. But I needed to find a club, so I ran to the side of the building.
I couldn't find one, Nell. I did…but it took awhile.
Returning to the fight…I entered it.
Nell, I don't know how much I helped.

Nell-

What happened?

Alf-

The bigger dog retreated…went over the gate.

Nell-

When Bobo saw you…he probably fought harder. If you were older, perhaps you would have injured the other dog…which would have been a mistake.

Alf-

Yes. You're right, Nell.

Nell-

Why did my sitting on your lap bring that scene?

Alf-

Some of my life flashed before me. I…saw a painful part of it…and a loving part. I've always had that scene in my heart.

Nell-

Al, I know how long it took…to find the stick.

Alf-

The reason Bobo whipped the Saint Bernard was to protect us. Sure! When my brother and I and Bobo walked, we were his territory. The Saint Bernard must have entered it, and started a fight.

Nell-

As you entered his place, maybe.

Alf-

Sure. And Bobo had us to defend.

Nell-

Some of our memories are painful…even as they are sweet imperfections.

Alf-

Yes, Nell.

Nell-

Unbutton my shirt, Al. See what that connects to.
See what that raises.

So he did. And...yes...it brought something up.

THE CONSCIOUSNESS FIELD

Part 2

Fate and Freewill

Alf and Nell have come into town, to Unalaska…often called Dutch Harbor. They're in a small house, and sit at a window table. The outer scene is of a misty day, and of a river you could throw a stone across.

On the other side of the river are two or three houses, nearby…these also small.

Nell-

Where's the wind, Al?
I like your friends. After we visit Nepal…shall we
come back?

Alf-

If we want to. I can work as a carpenter.

Nell-

Maybe you'll be selling books.
(she taps papers, on the table)
You're not pleased with your essay. Just a guess.

Alf-

This second part is about freewill and fate. Some of The thoughts are different from my other books. Are you in the mood to listen?

Nell-

I'm ready!

Alf (reads)-

We're imperfect! But in the lack…is love.
Consciousness is the basis of "Our Something", and it moves itself forward, as to its smallest parts.
(he looks at her)
Nell…a collection of words define conditions present as "Our Something" was enabled.
(he sighs)
"Possibility" was perhaps the first ethereality.
(he reads)
A oneness exists forward by enabling "twoness".
Movement can then serve the continuance. Thus… "change" occurred.

Nell-

Will you remind the reader the temperature was nearly zero? Tell them of studies showing that this would allow almost free flow of whatever energy radiates.
Sorry to interrupt.

Alf-

Consciousness radiated to "effect". An allowance of that was manifestation.
(he reads)

It became possible to go onward when "I" enabled "change".

Nell-

So…Al…what action is in the enablement? You say Consciousness is a field, both ethereal and manifest. And, somehow, it effects itself forward.
I remember you saying, of some interactions, that when a particle and its anti-particle collide, they are annihilated. Also…concerning the Higgs Boson, and what it encounters, there is effect through a change in movement, and it slows things.
So…what does Consciousness do to itself?
Certainly it doesn't annihilate.

Alf-

Nell…Consciousness is immediate. So…when it "encounters", the action enables a move back in time, and ahead. A connection occurs between the past part of the quanta unit, and the future part.

Nell-

You should remind the reader that things move forward in action at the smallest parts, the quanta.

Alf-

I will. Nell…do you believe in fate?

Nell-

Only if things are connected…and separate. You Called existence carpet-like.

Alf-

Nell, if "Our Something" is enabled by consciousness...this in the matter of laws, and the "now" of change in each quanta unit of itself, and the "I" we have, an ethereality not needing to be contained... all things in "Our Something" are Consciousness.

Nell-

Also...as you say, Al...there are universes. In ours, time and distance is one thing following another, as To what experience tells us is relative to the speed of light.

Alf-

Sure, and we must accept both realities. Here... tomorrow has not yet occurred and, yet it has, becausein all reality it has.

Nell-

Due to immediacy...throughout all reality?

Alf-

And "possibility", Nell...resident in freewill, even as to established laws.

Nell-

Al...I believe you, but the reader will be slow in accepting that.
(she looks through the window)
Our love is contained in our "I", which is free to choose...and to roam far.
We are living day to day...waiting for the sun to rise, and rise again.

When we perish, the sun will bid adieu…in its own journey.
You're making me sad, Al.

Alf-

Let me finish this, Nell…then we'll go to the hill and get blueberries, in the sun.
It's sad we're alone when we make choices.
Sure…how sweet to be a child again, holding the hand. But…no…

Nell-

But, yes, Al…for those in the flock.

Alf-

Some are there always, Nell, which is not the freedom you and I want. I shudder at the "always" part. Some things done to excess in that part are with us for centuries.

Nell-

I want to be fully human, Al…with freewill to choose at the bridge. I'll be alone there…erring often…but you can hold my hand.
(he holds her hand)
We have freewill available, Al.
Then…you made a case for the seer being able to see ahead. Yes, also, I agree to fate.
Many would agree to both, perhaps, if you got them to consider what you say.
Read from your papers.

Alf (reads)-

If freewill and fate are both true, there is our familiar time because tomorrow has not yet occurred, and zero time because it has.
To see this…it is best to consider that our familiar existence moves onward in the smallest units of itself, the quanta.
In every instant of change within the units, both times are represented.

Nell-

If in this instant I'm Nell Barton Jangmu Sherpa, I'm that in the next instant, but also different.
How does the quanta unit determine my identity and continuance?

Alf-

Nell, the "same as" would die if "different from" didn't simultaneously occur. The past and the future of the unit must be joined…far in excess of the speed of light…without annihilation.
You and I accept the continuance and change.
Consciousness moves all parts of itself onward by encounter.

Nell-

Al…each quanta unit has the past part, the "now" part…which exists as the ethereality of it…and the future part. When encounter occurs, immediacy brings a move back in time.
Also, according to Alfred…a move ahead in time.

Alf-

Yes, it's enabled that the past and future connect.
This is the "now" in the action, and we continually feel it.
We go on in the ethereality of "now".
Also, Nell...we go on in our "I", which is uncontained, or can be.

Nell-

It's in a containment. It is contained, Al. But, you say say it doesn't need to be.

Alf-

We go on due to Consciousness "becoming" in each unit. The times involved are our time and zero time.
(he slaps the table)
The manifest goes forward in "best path", according to the "now" in each movement.

Nell-

Al, I accept fate. It seems that all the "now's" are Consciousness. The carpet has the "same as" and "different from because...

Alf-

Because they're necessities of imperfection. You can call it love of existence.

Nell-

Let me continue, Al.
Although freewill is true, because we take one step after another...connected but separate...and have our

own choice as we act, man's "I" sometimes is cleared to the extent it can see ahead on the carpet. So…

Alf-

If man can't understand that both fate and freewill are true…he'll never fully accept responsibility for his actions. And…he won't see love as the basis of existence.

Nell, we ought to often ask, as did Chaucer, "How can the world be served?".

Nell-

But, Al, every time we act, we don't examine our ethics. We just need to have developed an inner decency.

And…we accept a lonely responsibility.

Alf-

Yes.

Nell-

Al…this paper is about fate and freewill. But you haven't given an example of it happening to you. Please tell us about Gladys.

Alf-

The working of fate? We all have encountered it, but sometimes it is dramatic and memorable.

Gladys…yes, I should speak of her.

I met her during my journey. I…

Nell-

What journey?

Alf-

The one we all take.
I had no Penelope, no kingdom aching for my return, no son among the robbers, no dog waiting for the sight of me.

Nell-

You were not Odysseus. You were Alfred...so...

Alf-

For awhile I was in the army of my country, and gone to far Deutschland. Then to school at the University of New Hampshire, and Columbia in Manhattan. I was at Boston's Mass. General Hospital, in the department which handles records.
Oh, I could tell some tales.
But...I won't.

Nell-

And so...what about Gladys?

Alf-

On the wind...I came to Ogunquit, Maine.

Nell-

I'm so glad you made it.

Alf-

Nell, I think of its sandy beach only as a strip, the forest being just there, with continuance of the solid...this of earth. Did you ever dig where granite tempers life?

Nell-

Yes, Al, in New Hampshire…on my journey. Then I departed…perhaps as not strong enough to move aside the controlling power and rest awhile.
And so…you came to Gladys.

Alf-

I needed a new place to stay.
It was down near the sea, a bit away from a rocky shore…and toward a docking area for boats.
It was an Inn. You might have thought of it as the former home of a wealthy ship's captain…now put forth to house the tourist.
And there was Gladys, alone in it.

Nell-

Will you describe the building, Al?

Alf-

It was yellow…with white trimming prominent… particularly so at the window shutters.
There were three levels.
Facing the road, just there…and the sea, which was a couple stone-throws away…the Inn had a white stairway up the front of it.
Nell, I arrived at Glady's place in late summer…a few weeks before leaves would begin changing colors in a dramatic fashion.
I went to the back of the building.

Nell-

Why?

Alf-

A note at the front door told me to do so.
In the back, my apartment, I soon learned, was at ground level. Steps beside its entrance led up to her place.

Nell-

What did she look like?

Alf-

Gladys had a nest of red hair. She was a heavy woman, a bit or two over five feet. So...not tall, or short...just round.
She had a medium-size dog, with white hair. He was puffy white. And, of course, he could speak. Which he was coaxed to do, one evening, in reference to a pot of beef stew she made for him.

Nell-

What did he say?

Alf-

Probably he said "Thank you, my dear...the stew was yummy."
Or...it could have been "Raaraaraa!"

Nell-

Thank you, Al.
Anyway...it seems that Gladys liked you.

Alf-

She would invite me upstairs...occasionally.
I lived there about a month. In mid-September the

wind moved me along…or, rather, back to New Hampshire, to the University, for graduate work. So…one evening in her kitchen, she invited me to sit.

Nell-

Al, sorry to interrupt, but the reader will wonder why you were in Ogunquit.
What were you up to?

Alf-

I came from Boston with friends wanting to rent a local farmhouse. After being with them a month or so…I moved into town.
I got a job with a landscaper…then knocked on Gladys' door.
Let's get on with this, Nell.
(taps his paper)

Nell-

One evening you were in the kitchen with Gladys.

Alf-

I was invited me to sit. It was a small table, and when she sat across from me, and asked to hold my hands, it was easily done. I mean…comfortably.
My hands were palm up, and she grasped them in a caressing way. Nell, it seemed to be a matter of her making contact…feeling my nature…absorbing my essence.
I'm glad to say that I was quiet...as she closed her eyes, and drifted. To what? Images?
She said "I see a misty place…and you live beside a river."

She was quiet for a moment. Then she said I would have a...uh...a something or other, with so and so. Of course, she did say a name.

Nell-

Who? What? A what...with who?

Alf-

It's over...it's...you know. It's past.
So and so, and I...knew each other awhile, and Gladys saw it clearly.

Nell-

Did her face turn red?

Alf-

And then, Nell...Gladys spoke of the mistiness, with me in it.
Remember I had no thought of moving to such a place. And when she said I'd be a fisherman, I put it aside as improbable.
Then she said what seemed ridiculous...that I'd be known far and wide in physics or chemistry or some such. Which thinking I figured to be wrong. It seems remarkable that I remember her saying it.

Nell-

You dobelieve Gladys saw ahead.

Alf-

Nell, it wasn't until years later that I arrived at an expanded interest in Consciousness. I didn't expect that to happen. The "far and wide", is you, probably.

Anyway, as Gladys held my hands…I had no feeling that she had a true gift. I only realized it when I was in the misty place she spoke of.

Nell-

You didn't believe her...about being a fisherman, or being known far and wide for having to do with physics, or some such. Did so and so believe it?

Alf-

Who?

Nell-

But…here you are, writing your fourth book about Consciousness. With me…"far and wide"…listening. I hate her!

Alf-

Who?

Nell-

Oh, Al…what else did Gladys say?

Alf-

As she was with the river scene, she said "I don't like what I see!"…and she was very troubled by the sight.
And…Nell…that's when I made a big mistake. I interrupted her.
I asked "Did someone die?" And I continued with two or three similar questions.
She looked toward me, but she was not present. Was not "here". So I probably made it difficult for her to be

here or there.
Nell, she was in a between thoughts condition, in which the seer is being of the "ahead" scene...and not connected to questions about it.

Nell-

You learned to be quiet...too late? Sorry!
She didn't like what she saw?

Alf-

She said "It's all right!"...in a calming way.
(he looks down at the table)
That's all, Nell. If I kept my mouth shut, she would have let me know of what she saw that she didn't like...and about other things.

Nell-

She did say "It's all right!". But, Al...what could she have seen that she didn't like?

Alf-

The important part would have been a specific truth. She might have seen any one of many bad happenings. We all have them, Nell.
I almost drowned when my kayak flipped in an offshore wind. My house was torn apart by a two hundred miler, as they nuclear blasted underground at Amchitka. Later, my redwood water-tank house burnt to the ground. I...

Nell-

Oh, my God! Al...

Alf-

I don't know that any of these happenings were part of what she referred to.

Nell-

What other events could have been seen?

Alf-

This essay isn't about me!

Nell (reaches across the table and holds his hands)-

Poor dear...to have such things happen.

Alf-

Nell, a seer could put forth "I don't like what I see!" to most hands held.
What of you?
Tell me!

Nell (closes her eyes, and rubs her forehead)-

When I was four...
(she sighs deeply)
My mother and I were on a trail northwest of Thame, and...yaks, laden with trading goods from Tibet, came along. My mother and I got against the wall on the inner part. There was no other thing to do...no way to climb higher, it being so steep...and no safe way to go lower, over that edge.
My mother knew the men in charge of the yaks. One should have been in front. She waved, and shouted... at the wrong time. Al, the lead animal thrust his head toward me, and his horn caught in my tunic.
When the head man yelled, the yak pulled away...but

I was yet caught, and he threw me over the edge.
(tears are in Nell's eyes)
Al, the yak didn't mean to do it.
I fell…in a bouncing way…not far…maybe thirty feet, to a ledge.
Below the ledge…was two thousand feet of air. I mean, in a straight down way.
(she looks at him)
I survived it!

Alf (holds her hands)-
I'm so glad!

Nell-
Also…I almost died just after I was born.
Well…here I am, and you, too.
Please read, Al.

Alf (picks up a paper)-
By thinking outward beyond this universe, and inward, to the smallest units of it, man can wonder about Consciousness as it relates to a larger part of "Our Something".
He might see that need, or love…defines the "onwardness" of it.
Consciousness is love…effecting continuation of all the parts.

Nell-
I'm so glad you spoke, earlier, of man needing to know he stands alone. The responsibility for choices he makes…is his only.
That's why he is human.

Alf-

Nell, if the "I" is messed with, manipulated, taken away…man's body and body-mind won't be responsive or responsible to what is fully human but, rather, to his genes, or to other men's directives.

Nell-

Al…from the time of Sophocles, and his writings, man has seen the importance of fate as a valid truth. But, without understanding that freewill dominates here, in this universe, man can lose his choices…his responsibility…and become robotic.
Aside from that…Al…when you say man stands alone, you will be opposed.
Doesn't God…or the "just there"…have to do with what happens?
This is what they will ask.

Alf-

All I can say, Nell…is that man has "I", and is free to make his choices.

Nell-

Al…you've admitted to knowing the feeling of being touched. Something "just there" was having to do with you.

Alf-

Nell…it is because we have "I", that we are free to destroy our possession of it.

Nell (sighs)-

I agree, Al.

Now…how can a seer see ahead? You've said she's
In a trance-like state, clearheaded…open. But how
can she be part of a scene which won't occur for
years to come? I know we must consider both views
of time and distance.
Tell us about your conclusions.

Alf (reads)-

As to "writtenness" for particulate things, movement
forward serves "best path". Laws were established to
allow this.
For us…Al and Nell…fate depends on freewill.
(he looks at Nell, then reads again)
Contained, yet unconfined…the human "I", as part of
all existence, can have "everywhere" available to it,
in the sense of contact with what is moving ahead,
and to its choosing to have it.
Gladys saw ahead relative to two aspects of reality,
The particulate and the ethereal.
(he looks up)
Nell, this is the year 2014.
Our "I" is connected through discriminatory consciousness…to this year's "now", as to a series of
personal connections.
The seer, in a trance-like state…between thoughts…
is free of being confined to the "now" of any series
within "change".
In Ogunquit, because Gladys was being of my
essence, perhaps due to holding my hands, she was
free to go ahead to dramatic results of my choices.
She went ahead to a misty place, where I live by a
river.

Nell-

She saw so and so…and you were encountering the hands of death. Two bitches. But…if love all things, you've got to love them, too.
Sorry!

Alf-

Yes. She saw entanglements deeply meaningful, as showing in the misty place. And she saw…that I would throw my thoughts into the wind.

Nell-

To have them carried to the four corners, Al.

Alf-

But, Nell, even a moron might succeed in that.

Nell-

Phhht!

Alf-

Nell…I've had my say about Consciousness.

Nell-

And if you didn't interrupt…Gladys would have seen me, standing in thin air.

Alf-

I saw you…and both of us have had a deep breath.

Nell-

So, Al…let's have a new climb. I'm ready!

AFTERWORD

Consciousness as the Primary Field

Alf-

Nell, do you think every existent was preceded by the possibility it could exist?

Nell-

My logic says “Yes”. That’s why I think existence didn’t begin with this universe. But…

Alf-

I call the larger reality “Our Something”. The entirety of it is a field…the Consciousness Field.
It has particle-like action which allows manifest parts to self-maintain forward.
Nell, Consciousness is immediate…and so is its action, as it encounters.

Nell-

And…?

Alf-

If we eliminate barriers assumed in our relation to light’s speed, we can have clearer understanding of cause and effect, the separable nature of things, freewill, and fate.

Made in United States
North Haven, CT
01 December 2023

44858232R00150